W0081989

Rethinking Human Enhancement

Also by Laura Y. Cabrera

NANOTECHNOLOGY: Beyond Human Nature?

Rethinking Human Enhancement

Social Enhancement and Emergent Technologies

Laura Y. Cabrera
University of British Columbia, Canada

© Laura Y. Cabrera 2015
Foreword © Steve Fuller 2015

All rights reserved. No reproduction, copy or transmission of this publication may be made without written permission.

No portion of this publication may be reproduced, copied or transmitted save with written permission or in accordance with the provisions of the Copyright, Designs and Patents Act 1988, or under the terms of any licence permitting limited copying issued by the Copyright Licensing Agency, Saffron House, 6–10 Kirby Street, London EC1N 8TS.

Any person who does any unauthorized act in relation to this publication may be liable to criminal prosecution and civil claims for damages.

The author has asserted her right to be identified as the author of this work in accordance with the Copyright, Designs and Patents Act 1988.

First published 2015 by
PALGRAVE MACMILLAN

Palgrave Macmillan in the UK is an imprint of Macmillan Publishers Limited, registered in England, company number 785998, of Houndmills, Basingstoke, Hampshire RG21 6XS.

Palgrave Macmillan in the US is a division of St Martin's Press LLC, 175 Fifth Avenue, New York, NY 10010.

Palgrave Macmillan is the global academic imprint of the above companies and has companies and representatives throughout the world.

Palgrave® and Macmillan® are registered trademarks in the United States, the United Kingdom, Europe and other countries.

ISBN 978–1–137–40223–3

This book is printed on paper suitable for recycling and made from fully managed and sustained forest sources. Logging, pulping and manufacturing processes are expected to conform to the environmental regulations of the country of origin.

A catalogue record for this book is available from the British Library.

Library of Congress Cataloging-in-Publication Data
Cabrera, Laura Y., 1983–
    Rethinking human enhancement: social enhancement and emergent
    technologies / Laura Y. Cabrera, University of British Columbia, Canada.
        pages   cm
    Includes bibliographical references.
    ISBN 978–1–137–40223–3
    1. Prosthesis—Social aspects.   2. Biotechnology—Social aspects.
    3. Socialization.   4. Self-help devices for people with disabilities.
    5. Well-being.   I. Title.
    RD130.C335 2015
    617.9—dc23   2015012897

To Mattias, the light of my life

Contents

Tables

Foreword

I first encountered Laura Cabrera at a conference organised by Christopher Coenen at Karlsruhe Institute of Technology, Germany, in July 2013, where she spoke very eloquently about the need for 'visioneering' in social studies of science and technology. On this basis I asked her to join the Social Epistemology Collective. 'Visioneering' is a Silicon Valley neologism combining 'vision' and 'engineering'. It is meant to capture people who are trying to realise their vision of the future. To the cynical this looks like a rebranding of 'entrepreneurship' for potential wealth producers who do not come from capitalism's traditional industrial base. And that may well be true—but it is not the entire story. Much more interesting is the extent to which visioneers are willing to disrupt default market dynamics. It is in this proactionary spirit that the reader will appreciate Cabrera's *Rethinking Human Enhancement*.

Cabrera's perspective is distinctive in treating the prospects for human enhancement in a policy-friendly manner. She spells out three enhancement paradigms in terms of philosophical motivation, evidentiary base, promised outcomes and likely consequences. Moreover, she comparatively evaluates them, arriving at an ordering of policy priorities. In pole position is what Cabrera calls 'social enhancement', followed by enhancements to individuals based on the extension of current biomedical knowledge, with the more cyborg-like 'transhumanist' enhancements bringing up the rear. Although she presents this order as proceeding in terms of increasing risk to the sustainability of the human condition, strictly speaking it is about increasing intervention into the default processes of the human organism. Here it is worth saying that although Cabrera accords the lowest priority to the core transhumanist aim of 'morphological freedom' (i.e. the freedom to change one's own body with impunity), hers is one of the few policy frameworks that actually grants it any serious standing.

As with many contemporary proposals for improving the human condition, social enhancement's ideas have been with us before, typically in cruder guises and scarier language. Just as the prehistory of biomedical enhancement lies in eugenics, the prehistory of social enhancement reaches back to the design of 'smart environments', as the cognitive scientist Donald Norman started calling them in the 1980s, which in turn incorporated ideas from the 1950s and the 1960s of operant conditioning and biofeedback. Students of this line of thought will recognise the trajectory as starting from the need to take full advantage of the individual's capacities in situations where more conventional forms of training won't work because the information that the individual needs to have for an adequate response is too context-dependent. Such are the open-ended situations thrown up in times of war or possible war, the original context for this research.

However, by the time Norman's 'smart environment' idea started taking hold, the Cold War was winding down and—as typically happens to research developed originally for military purposes—its projects were being redeployed for civilian purposes. The generic use of the word 'dashboard' to cover the control interface of 'smart devices' (i.e. devices sensitive to the user's particular needs and tendencies) is a legacy of this redeployment. But in our own times, interest in human enhancement has begun to return to the public policy agenda, as demonstrated by 'converging technologies' initiatives on both sides of the Atlantic, which provide much of the pretext for Cabrera's study. As its European proponents have made explicit, this development has reflected the need to render welfare state citizens more efficient producers and consumers— and for a longer time—in order to deliver on the promises of social justice in times when public resources are limited if not in deficit.

The reception of the radical behaviourist B. F. Skinner in the third quarter of the twentieth century provides a precedent for the uphill public relations battle that 'social enhancement' is likely to face. Skinner's ideas were widely feared and loathed, even when presented as an attractive piece of fiction, such as *Walden Two*. In the freedom-loving West, they were seen as delivering a sugar-coated version of the worst features of Soviet-style social engineering. What made Skinner seem so pernicious was that his version of behaviourism was founded on the principle that you can get an organism to do virtually anything if you tie it to something that the environment already rewards the organism for doing. In that case, the organism can learn to do what you want as part of what it normally does. It is a 'smart environment' in the most literal sense, where the environment unobtrusively smartens up the organism according to the designer's standards of intelligence.

However, the 'intelligent designer' is absent from the proceedings, since both the stimuli and the responses are self-administered. Such smart environments were originally dubbed 'Skinner boxes', in effect a technological micro-realisation of Deism. Consider the amount of information that people routinely divulge about themselves in order to access facilities in cyberspace. The 'surplus value' generated by such internet-based transactions is then sold as a commodity to companies wishing to market their goods to the users of those facilities. But even this outcome is relatively modest in terms of Skinner's original aspirations. A more natural heir to his radical behaviourism is the emerging phenomenon of 'serious gaming', whereby video games are used as platforms for reorienting attitudes and improving capacities well beyond the confines of personal entertainment. (The most thoughtful popularisation of this project is probably Jane McGonigal's *Reality Is Broken*.) The value that the parties themselves derive from flourishing in these smart environments—often described in terms of 'fun'—is relatively independent of the system-level benefits of their activities.

Indeed, this was the original selling point of Skinner's operant conditioning paradigm among fellow behaviourists: efficiency. The behaviourist did not need to inform subjects at regular intervals whether they responded properly. Subjects would learn that for themselves over time, with the behaviourist manipulating their 'reinforcement schedules' from behind the scenes to increase their response efficiency. Today's behavioural economics (aka 'nudge') owes much to Skinner's way of thinking, especially its preference for manipulating the information that is made available to subjects (with subjects then let to decide for themselves) over more explicit strategies of changing their behaviour, such as financial incentives and coercive legislation. Little surprise, then, that despite people's misgivings Skinner was the most widely cited psychologist after Freud in my youth, 40 years ago— but that was because his smart environments were regarded as more inevitable than welcomed.

But Skinner pushed the point still further. In his 1971 bestseller *Beyond Freedom and Dignity*, he spelled out the philosophical implications of his proposals—in particular, that 'free will' and 'consciousness' are no more than decision-making platforms that happen to be located inside our bodies but might as well be on a dashboard, in today's lingo. It was a direct attack on those who located our 'humanity' in some inner life that is in principle unavailable for public inspection. As a result, Skinner was seen as 'de-humanising' us, a charge he accepted with equanimity because, like a lot of Protestants who question the faith of those unwilling to declare it publicly, he wondered whether beneath the variability of human response really lies an 'inner life' that resists public expression. Put bluntly, Skinner was inclined to diagnose the 'mystery of consciousness' as symptomatic of either cowardice or deception—if not sheer ignorance of what one is capable of doing under the right circumstances. And whatever else one wishes to say about Skinner here, his sense of the inquisitorial power of the scientific method was of a pure sort that hadn't been seen since the days of its lawyer-founder, Francis Bacon.

Nowadays, in a triumph of intellectual public relations, we speak of the 'extended mind' and 'distributed cognition' to paper over Skinner's philosophical indiscretions. Thus, instead of claiming that we have lost our minds to environmental contingencies that reinforce certain behaviours, we now say that our minds are partly in our bodies – but also partly, if not mostly, outside our bodies in the things that enable us to be as intelligent as we wish to credit ourselves. To put the matter in brutally physicalistic terms, rather than simply eliminating the very idea of mind, as Skinner would have done, we spread the idea out as thinly as possible. But just how thinly can we spread it out—and how evenly? After all, even the most ardent devotees of the world view described here—say, Bruno Latour or Andy Clark—do not assume that things outside the human body that contribute to 'human intelligence' contribute to the same extent as each other, let alone to the same extent as the human body itself.

But this then opens up interesting political and legal issues: who or what takes responsibility for the course of events, given what is understood to be the power relations at play in enabling various expressions of 'human' intelligence? However one answers this question, the idea of 'social enhancement' is likely to focus them more effectively than that of 'moral enhancement', a phrase that has featured prominently in recent analytic bioethics, largely thanks to Peter Singer's student Julian Savulescu. Moral enhancement starts by assuming, on what is advertised as evolutionary grounds, that humans are capable of inflicting more harm than good on each other in the normal course of things, a situation that needs to be rectified to enable humans to live together in relative harmony. To be sure, the start is wrong-footed in discounting the benefits that come from our normally dealing with each other without a looming sense of potential harm, even when the likely outcomes are not directly beneficial. However, moral enhancers are more charitably understood as offering a cognitivist twist on the idea that humans are self-interested individuals: Instead of the overweening nature of human desires, the stress is placed on the damage caused to others in their pursuit. As a result, the need for moral enhancement is presented as a disease that needs to be cured, rather than a corrective to a largely successful approach to life, which was how classical liberalism tended to treat matters.

But more to the point for Cabrera, the moral enhancer's level of intervention for remedying the situation is the individual. Thus, much of the moral enhancement literature is devoted to the ethics of giving people drugs to become, say, altruistic—as opposed to allowing them to discover altruism in the normal course of maturing into an adult. (Here readers might appreciate Cabrera's ironic re-appropriation of 'HE'—which normally stands for 'higher education'—to mean 'human enhancement'.) Let us set aside the obvious objection that based on what we know about both genetics and neuroscience, it is unlikely that any biochemical interventions would produce exactly the effect we want—and nothing else that might undercut the anticipated benefits. We are still left with a highly individualistic way of thinking about the human condition, which boils down to the equation, 'good individuals = good society'. At this point, Skinner kicks in—and the idea of social enhancement looks more plausible. (But this is not to deny that cash-strapped welfare states might be tempted to try drugs as a cheap replacement for 'socialisation'.)

In order to have a society that maximises benefits to the most people while at the same time allowing each individual the greatest level of personal autonomy (i.e. the Benthamite and Kantian imperatives, respectively), it is not necessary that people act 'morally' in some conventional sense. A moral attribute such as altruism may be a systemic virtue of society but not a virtue of its members. In that case, altruism would be an emergent consequence of a process that enables certain forms of social interaction which capitalise on individuals' default tendencies to produce a desired result: a smart environment designed to maximise social enhancement. This is the normative space—previously occupied by theodicy in Christian theology—that Cabrera's work occupies. In today's world, it would also serve as a great launch pad for reviving the sense of scientific enquiry that originally inspired Francis Bacon.

Steve Fuller
Auguste Comte Professor of Social Epistemology
University of Warwick

Preface

During the last decade human enhancement has become a much-debated topic. Different values and visions underlying our understanding of human enhancement can have a significant impact on the way emergent technologies are used to improve the human condition. In view of this, there is a need to explore ways in which emergent technologies can be used in a pragmatic, feasible and more ethical way for improving the human condition. The aim of this book is to contribute and bring new insights to this quest. Towards this end, the features of this book include:

- *An extensive introduction providing an overview of human enhancement and the role of emergent technologies in this human quest.* The introduction will provide the reader with enough background information to be able to proceed to the following chapters. In addition the introduction also provides an overview of key emergent technologies.
- *Discussion of three human enhancement paradigms.* Three different understandings of human enhancement are discussed, each one in a separate chapter. Such a categorisation helps to capture the different views that have been emerging or that have been neglected around the human enhancement discourse and from which a more fruitful analysis of the ethical and societal issues involved can take place. The two predominant views of human enhancement, namely the biomedical and the transhumanist, are analysed first. Both of these understandings hold a view about the individual as isolated and abstract, which has led to the use of technologies with a focus on individualistic types of interventions, in particular interventions aimed at directly changing the individual's body and mental features. The biomedical understanding is oriented towards developing or improving medical-related interventions, whereas the transhumanist is focused on developing radical and often disruptive high-technological interventions, which often are risky and expensive. After exploring these particular perspectives on human enhancement and the role technology is likely to play in each one of them, a third paradigm is introduced. This third paradigm is referred to as social enhancement, and is characterised by a relational view about the individual. This is an alternative way in which to

understand human enhancement. One that shifts the focus of attention to social- and community-oriented interventions, promoting the use of emergent technologies for environmental enhancement interventions that are safe and accessible, rather than interventions focused on changing the individual directly.

- *Framework for the prioritisation of human enhancement applications.* This book discusses some of the most pressing, but not necessarily obvious ethical issues of human enhancement and offers as a suggestion a prioritisation scheme focused on the particular uses of emergent technology for human enhancement. In particular it puts forward that we have morally compelling reasons to promote a framework in which social enhancement is prioritised over the other enhancement paradigms. This leaves biomedical enhancement interventions as a second-best option followed by those enhancement interventions suggested by the transhumanist paradigm.

The first chapter introduces and sets the context and background for the topic that this book deals with, namely the ethical considerations for choosing one kind of enhancement over another when using certain technologies.

Chapter 2 introduces the first human enhancement paradigm to be analysed, the biomedical paradigm—based on the therapy-enhancement distinction—which is taken to be the current and dominant one. The first part of this chapter starts by describing some key concepts in which the biomedical understanding of human enhancement is based, such as health, disease and therapy, and then goes on to describe the main features of the paradigm. The second part of the chapter explores the ethics of the paradigm, by first exposing its stand around the main arguments concerning enhancement, second analysing some of its pitfalls and finally mentioning the valuable things this paradigm brings to the fore.

Chapter 3 addresses the transhumanist paradigm of human enhancement, which is one that is gaining more support owing to the strongly emerging transhumanist movement. This chapter follows the same structure as the previous one. The first part describes some key ideas that have motivated the transhumanist understanding of human enhancement, such as transhumanism, posthumanism and the posthuman, and then describes the main features of the paradigm. The second part goes on to explore the ethics of the paradigm.

Chapter 4 puts forward the third paradigm of human enhancement, the social paradigm, which is a neglected but valuable human enhancement understanding. Following a similar structure to chapters 3 and 4, the chapter explores whether or not this understanding brings something new to the discussion on human enhancement and then puts forward the key ideas motivating this paradigm—social justice and social determinant approaches. The main features of the paradigm are also described. The second part of the chapter goes on to explore the ethics of the paradigm, by analysing from its perspective its position on the main arguments around enhancement, some of its pitfalls and its potential value.

Chapter 5 questions if the real dilemma around enhancement is to enhance or not to enhance. The first part of the chapter focuses on issues that are more salient to brain enhancement interventions, while the second part discusses other ethical concerns that have not been given sufficient consideration by previous researchers or have not been covered as widely as those mentioned in the different human enhancement paradigms.

Chapter 6 argues that given the different features each paradigm has, the different ethical issues each paradigm instantiates, and considering the pressing global problems we are facing, we have morally compelling reasons to prioritise enhancement interventions according to the social enhancement paradigm. We then can focus on those interventions suggested by the biomedical one, and finally on those suggested by the transhumanist paradigm. The final chapter revisits some of major issues discussed in this work and sketches some recommendations for moving ahead.

This is a book of enquiry in an emerging field. As such it does not aim to provide all the answers to the enhancement debate, but it hopes to provide a few insights, and provoke thought and debate that may lead to a deeper understanding of the role human enhancement and emergent technologies can and should play in society and for individuals. Hence the contribution of this book is confined to shedding light on issues to which further consideration needs to be given in order to address the different challenges with which human enhancement confronts us.

Acknowledgements

I would like first to thank my parents (Israel Cabrera Vargas and Laura Trujillo Salinas) and my husband Mattias Olauson for their support and love. I would also like to give a special acknowledgement to my supervisor Professor John Weckert for his comments, and for guiding me in the process of my work. I would also like to give a special thanks to Professor Steve Fuller for encouraging me to publish this book.

There are many more who helped me shape my ideas and thoughts to whom I am in debt for their challenging questions, comments and suggestions.

Abbreviations

AI	Artificial Intelligence
BCI	Brain Computer Interface
CSDH	Commission on Social Determinants of Health
CRNNI	Committee to Review the National Nanotechnology Initiative
DARPA	Defence Advanced Research Projects Agency
DBS	Deep Brain Stimulation
DNA	Deoxyribonucleic Acid
EOC	Evolutionary Optimality Challenge
ETC	Erosion, Technology and Concentration
EURON	European Robotics Research Network
fMRI	functional Magnetic Resonance Imaging
GRAIN	Genetics, Robotics, Artificial Intelligence and Nanotechnology
HLEG	High Level Expert Group
ICT	Information and Communication Technologies
IEEE	Institute of Electrical and Electronics Engineers
IERC	European Research Cluster on the Internet of Things
ISO	International Organization for Standardization
ITC	Information Technology and Communication
LOC	Lab-on-a-chip
MEG	Magnetoencephalography
NBIC	Nanotechnology, Biotechnology, Information Technology and Cognitive Science
NIO	Neurotechnology Industry Organization
NNTI	National Neurotechnology Initiative
NRC	National Research Council
NSTC	National Science and Technology Council
PCB	President's Council on Bioethics

PEN	Project on Emerging Nanotechnologies
PCSBI	Presidential Commission for the Study of Bioethical Issues
R&D	Research and Development
SDH	Social Determinant of Health
tDCS	transcranial Direct Current Stimulation
TMS	Transcranial Magnetic Stimulation
UNESCO	United Nations Educational, Scientific and Cultural Organization
WHO	World Health Organization
WTA	World Transhumanist Association

1
Introduction to the Enhancement Debate

Background

'Human enhancement' has become an umbrella term to refer to a wide range of existing, emerging and visionary technological interventions that blur the boundaries between interventions aimed at therapy and those beyond therapy, as well as interventions aimed at prevention, restoration, rehabilitation and promotion of well-being. Discussion of issues related to human enhancement is nothing new. The human desire for improvement goes before we have even developed any sophisticated technology. History has shown us that the desire for more, for the unlimited, for better and for the different is not satisfied with the average, nor takes its weight from the distinction between healthy and better than healthy or the abnormal and normal. In this sense, Bertrand Russell was probably right when stating that humans differ from other animals, insofar as they have some desires that are never satisfied. We can even say that the desire to improve ourselves and to overcome our limitations is all-too-human.

Throughout history, understanding of the human condition and ways in which to improve it have changed as a result of the various interactions we, as humankind, have with technology and our environments; changing the way in which we perceive and value our world and ourselves. The dynamic process of trying to improve the human condition is shaped by social changes but also by technological and scientific developments.[1] It is through science that we get a better understanding of our environment, and with technology we develop the tools and devices that day by day help humankind to understand his condition and possible ways to improve it. Thus, even though the desire to improve ourselves has been part of humans for a long time, advances in emergent technologies will likely provide us with unprecedented ways to improve the human condition. While these technologies have the potential to change the human condition for the better, they also have the potential to change it for the worse.

Over the last three and half decades, advances in different technologies—such as pharmaceutical, genetic interventions, implant technology—have given a particular twist to the human desire for improvement (Bostrom & Savulescu, 2009; Buchanan, Brock, Daniels, & Wikler, 2001; Coenen et al., 2009; Daniels, 1992, 2000; Elliott, 2004; Juengst, 1997, 1998; Lin & Allhoff, 2006; Parens, 1998, 2006; PCB, 2003; Savulescu, Meulen, & Kahane, 2011; Williams, 2007). In the literature the tools, methods or substances used for the purposes of enhancing interventions are referred to as 'human enhancement technologies' (Coenen et al., 2009; Elliott, 1998, 2004).

Human enhancement interventions are not confined to medical practice. Military applications and entertainment are examples of other fields where human enhancement interventions are used. The discussions around human enhancement have not only become a fashionable topic in certain circles, but the literature on it has reached a critical mass, which makes it a major topic of ethical research (Bostrom & Savulescu, 2009; Coenen et al., 2009; Selgelid, 2009). Apart from human enhancement being an all-too-human desire, other reasons for it being such a hot topic are the current cultural, economic and political dynamics that are pushing the rapid development of human enhancement technologies—such as national security concerns, global competitiveness, brain drain and consumerist lifestyle demands. Given the pervasiveness and the impact of the topic, it is not far-fetched to say that we are indeed near the start of a human enhancement revolution (Allhoff, Lin, Moor, & Weckert, 2010; Lin & Allhoff, 2008b) or that we live in an 'enhancement society' (Coenen et al., 2009). That is why the debate on technology and human enhancement has been regarded as one of the most crucial debates of our time.

The human enhancement discourse touches upon a wide range of concerns, from philosophical and ethical approaches to the proper role of medicine and desirable human qualities (Wolpe, 2002). Human enhancement has been portrayed as a liberating and exciting opportunity, or as a disturbing and even frightening one, that is why it carries both the promise of 'superhumanization' as well as 'dehumanization' (PCB, 2003). The term 'enhancement' by itself is already problematic. Many people regard it as abstract and imprecise, as it can be understood in many ways. When we talk about enhancing a function, a capacity or a trait, are we referring to making more of it or making it better? If we are referring to the former, what standard are we supposed to take as a base? In the human case, saying that something is better can mean better than I have *done before*, or better than my *opponent* or better than *the best* (that is to say, than any other human being before). Enhancement can also mean bringing out a function or condition more fully rather than altering it qualitatively, or augmenting or improving a function or condition. All these differences are reflected in the various definitions found in the literature.

Generally speaking, the term 'human enhancement' is used to refer to any intervention or activity by which we improve or augment in any sense (e.g. performance, appearance) our abilities, bodies, minds and well-being (Lin & Allhoff, 2008b; Miller & Wilsdon, 2006; Williams, 2007). Thus, we can think of our daily life being impregnated with enhancement practices—such as exercising, drinking coffee, having a good diet, taking vitamin supplements and education. These examples can be regarded as 'natural' human enhancement interventions, which tend to be seen as not morally interesting and unproblematic (Allhoff et al., 2010; Coenen et al., 2009), where the improvement attained is usually not perceived as an enhancement, even though by definition it is. For instance, people generally do not make a big deal out of someone drinking coffee in the morning in order to improve his or her performance in the office, or of someone studying in the best-ranking university rather than a small public school.

It has been argued that if we were to take such a broad view on enhancement, where nearly everything we do could be considered as an enhancement, it would make the concept impotent (Bostrom & Savulescu, 2009; Coenen et al., 2009). The challenge has been to find a more 'reasonably intelligible' and 'non-arbitrary' account for it. So even if the term 'enhancement' remains somehow vague we still need an account that captures 'something that might plausibly be thought of as a kind' (Bostrom & Savulescu, 2009, p. 3); more importantly, an account that help us track morally relevant distinctions. A first step to achieve such a view has been to keep the discourse on human enhancement focused on improvements or augmentations brought about by *science-based* or *technology-based* interventions, such as those envisioned and suggested in the emergent technologies discourse. It is primarily these kinds of interventions that have instantiated ethical debate and controversy. Take, for example, the case of brain-machine interfaces. Their use by impaired people in order to enable them to achieve certain tasks seems uncontroversial; however, the reaction is not always of acceptance when brain-machine interfaces are aimed at other purposes, such as to augment performance in *healthy individuals*.

The problem with such an understanding of human enhancement is that, in a way, all technology is a form of human enhancement (Harris, 2007), as it improves our native human capacities, 'enabling us to achieve certain effects that would otherwise require more effort or be altogether beyond our power' (Bostrom & Savulescu, 2009, p. 2). In order to avoid this problem, some scholars have emphasised that we should focus on human enhancement interventions 'in the human body' (Coenen et al., 2009, p. 6). However, even if we were to agree on this point, just considering interventions *in the human body*, ambiguity remains regarding whether by this we mean enhancement interventions that are internal (inside the body) or external to the body (such as a prosthetic hand). It might also be needed to differentiate between interventions that have long-term effects or permanent results and those that have just temporary or reversible results (Bruce, 2007). Others have suggested that we need to differentiate between enhancement interventions that involve a radical change of state regarding human capabilities compared to those that only offer a change in degree in the expansion of human capabilities or power (e.g. incremental enhancement) (Bruce, 2007; Khushf, 2005). Some others have suggested that it can be helpful to distinguish between the ends—enhancing end states—and the means used to enhance (Kamm, 2009). Finally others have argued that we need to separate enhancement interventions that offer positional or competitive advantages (such as an increase in height) from those that do not (such as a better immune system) (Bostrom, 2005a; Buchanan et al., 2001; Elliott, 1998).

There are of course many problems in trying to justify why such distinctions would be meaningful. For instance, why should having a better immune system not count as offering a positional advantage? Better health generally means better life opportunities. Thus, it can be argued that any enhancement connotes in a way a positional advantage; otherwise it would not be regarded as an enhancement in the first place. Or how is it that taking coffee to keep awake is any different from taking a medically tested and safe drug with the same effect? Moreover, we can ask, if the underlying idea of human enhancement interventions is self-improvement and well-being, why is it that some enhancement interventions should be permissible but not others? Should human enhancement be limited to eliminating suffering and sickness, or should it cover the improvement of other goods such as intelligence, memory or beauty?

Considering all this, it can be argued that finding a meaningful and politically viable way to talk about human enhancement, not only for deliberating about the acceptable and the unacceptable, but more importantly to grasp the value of it, becomes a priority. This takes us to the core issue to be tackled in this book, the role of sound ethical considerations in the decisions we make over the kind of enhancement interventions we pursue.

Statement of the problem

Science and technology have challenged our dominating views of the world. First the Copernican revolution, which displaced the belief that our home planet was the centre of the universe; then Darwin dispelled the belief that man was divine, relegating us to descendants from the animal world; and later Freud demonstrated that we are not even masters in our own house. Now adding to the cosmological, biological and psychological shocks to human pride, we face the challenge of finding a legitimate distinction between us and our technologies (Mazlish, 1967).[2] Vernor Vinge predicted in his paper 'Technological Singularity' (1993) that the human era will end once we create superhuman intelligence. More recently, Raymond Kurzweil argued in his book *The Singularity Is Near* that we are already reaching that point, and that it is happening at an exponential speed rate (2005). Other authors, the majority of whom support the transhumanist movement, who have also put forward reasons for thinking that this could happen within the first half of this century, include Marvin Minsky, Hans Moravec and Nick Bostrom. They named this new era, in which our biological portion will become obsolete, the posthuman era (Bostrom, 2005a). These views help explain why the prospect of using different technologies for human enhancement has been the subject of considerable scientific research, and recently also of ethical debate.

With our rapidly evolving technology, we have also accelerated the rate of change affecting the human condition. According to some futurists, such as Kurzweil (2005), our technological developments follow an exponential accelerating rate, in which the number of technological breakthroughs has increased rapidly in comparison with the technological breakthroughs we have had so far in human history. However, it is questionable that our wisdom in using these developments in pragmatic and ethical ways in the quest for improving the human condition has followed a similar path.

For instance, considering the potential impact of emergent technologies, as well as the large amount of money available for research and development of emergent technologies in recent years, it is fair to say that there is still much to be done around the ethical, legal and societal implications that these technologies can give rise to. Moreover, there has not been much done in trying to find ways to use human enhancement and new technologies in less individualistic and more community-based ways.[3]

Even though the importance of ethical discussion is recognised, little has been done to foster it, and more troublesome is the fact that when ethical discussion does take place it often limits itself to problematic assumptions, not only around enhancement but also about the individual and the role of ethics itself. Public fear, rejection, great expectations or even unreasonable beliefs about emergent technologies can be the result of a lack of dialogue between governments, research institutes, granting bodies, researchers and the public on the implications and directions of these technologies (Mnyusiwalla, Daar, & Singer, 2003). That is why it is of great importance to have an adequate study of the ethical and societal implications of emergent technologies; in particular those aimed at human enhancement applications.

Under the dominant view of human enhancement—the biomedical one—emergent technologies are used to augment or increase our capacities (such as increasing our mental capacity or our sensory system limits) beyond any therapeutic aim yet still within the range of possible human species features. The transhumanist movement, which is becoming popular and influencing social views, is promoting a view in which emergent technologies are envisioned as a path to achieve the posthuman; that is to say, the augmentation of our human capabilities and features beyond those known to be possible for the human species. Both the biomedical and the transhumanist views of enhancement suggest a path of action aimed at directly manipulating the individual, whether it is the brain or its senses or any other aspect of an individual's biology. Thus, one can say that they are both focused on individual and individualistic bodily oriented enhancement interventions.

Focusing the discussion on individualistic, bodily oriented and fiction-like human enhancement interventions tells us something about the dominant values and views about what constitutes well-being. For instance, one may question why it is that most human enhancement applications that have been debated to date are those focused mainly on changing the individual's body or mind in a direct (and sometimes permanent) way.

Considering the kind of enhancement interventions that advances in emergent technologies could potentially bring to the fore, a more pragmatic, politically feasible and more ethical way to frame human enhancement is required. With this in mind, as an alternative way in which human enhancement can be conceptualised beyond the individualistic, bodily oriented and fiction-like kinds of enhancement interventions, a third view of enhancement is suggested: the social.

Considering the three different human enhancement paradigms put forward in this book—biomedical, transhumanist and social—the general question to be explored is:

> What are the ethical considerations for choosing one kind of enhancement over another?

This book's main argument is that given the degree and kind of ethical considerations involved in each human enhancement paradigm, we have morally compelling reasons to opt to use advanced technologies in different ways from those suggested by current human enhancement understanding.

Aim and scope

The aim of this book is to analyse and put forward a position on issues that have been latent for a long time in the fields of biotechnology, pharmaceutics and genetics, and have been brought to the fore again with more novel forms of emergent technologies, namely issues related to human enhancement. In the hope of adding something valuable to the debate on human enhancement, this work discusses three possible human enhancement paradigms and explores how each paradigm involves different values and different uses of emergent technologies, as well as different degrees and in some cases different kinds of ethical issues.

In the past there have been attempts to find alternatives to the therapy-enhancement distinction, for example:

- Gregor Wolbring has suggested a distinction between therapeutic versus non-therapeutic enhancement interventions (Wolbring, 2005). A recent European report has followed a similar line suggesting a 'non-medical' three-level distinction of enhancement:[4] (1) restorative or preventive non-enhancing interventions, (2) therapeutic enhancement and (3) non-therapeutic enhancement interventions (Coenen et al., 2009).
- Others have suggested distinguishing between enhancement and (1) augmentation, (2) alteration (Jotterand, 2008), (3) design evolution (Canton, 2004) or (4) human reengineering (Hook, 2007).

- Ruth Chadwick has suggested four different ways in which enhancement may be understood: (1) beyond therapy, (2) quantitative view, (3) qualitative view (enhancement different to improvement) and (4) as an umbrella term (Chadwick, 2009).

- Julian Savulescu has suggested at least two other main approaches to enhancement in addition to the 'not medicine' approach to enhancement (that is to say, treatment versus enhancement), the functional-augmentative approach and the welfarist approach (Savulescu, 2006). The former considers interventions as enhancements 'insofar as they improve some capacity or function by *increasing* the ability of the function to do what it normally does' (Earp, Sandberg, Kahane, & Savulescu, 2014, p. 2), whereas the latter considers changes to the biology or psychology of a person as enhancements insofar as they increase the chances of leading a good life in a given set of circumstances.

- Others have suggested dropping the term 'enhancement' because it is already politically charged in both its use and meaning. Zack Lynch, executive director of the Neurosociety Institute, has for instance recommended replacing it with the term 'enablement' (Williams, 2007).

- Finally, one more suggestion argues that instead of talking about therapy and enhancement 'as though these were categorically different things' and abandoning the therapy-enhancement distinction, we should focus on particular interventions, 'and examine the ethics of these on a case by case basis' (Selgelid, 2007, p. 1; c.f. Coenen et al., 2009), classifying them in terms of degree and prototypical cases. While such an approach might indeed be a valuable and reasonable one, it still faces the challenge of how different cases can be defined. This is often known as the problem of the *relevant description* (Daniels, 1992, 2008; Elliott, 1998).

Is a case defined according to the type of enhancement intervention? If yes, the problem would be that certain enhancement interventions while indeed an enhancement for certain individuals, might turn out to be quite the opposite for others. Let us consider the case of suppressing the ε4 allele of apolipoprotein E. There is evidence that having one or two copies of this allele increases the risk of developing Alzheimer's disease while at the same time it lowers the incidence of childhood diarrhoea and may protect cognitive development (Oriá et al., 2005). Thus the suppression of this allele can be considered an enhancement for those individuals living in healthy and supportive environments, while it can be a disadvantage for those individuals growing up in poverty and poor sanitation environments.

Is a case defined by the particular technology used? If yes, it would mean that the enhancement itself would not be as problematic as the means used to attain it. Thus, coming back to the allele example, it would mean that suppressing the allele would count as a different case if it was done through genetic engineering than if it was done through nanotechnology-based pharmaceuticals.

Is a case defined by a particular individual intervention? If yes, then the approach would require too much time and money, making it unappealing and not too practical. However, it could also allow us to consider in our assessment the motivations and desires underlying each individual case. To make this point clear, consider the example given by Carl Elliot (1998). Elliot invites us to compare the case of someone undergoing hormone growth intervention as a result of having a hormone deficit with a second case in which the individual seeks the intervention because he or she has a shorter stature compared with the rest of his or her community as a result of inheritance. Elliot invites the reader to consider how these two cases are different if the net result would be the same, namely the increase in height of the respective individual. Why would it be appropriate in one case and not in the other? The biological reasons are indeed different, and while in both cases the underlying reason might be that the individuals do not wish to be disadvantaged owing to their particular stature, the justification each individual has is not the same. In the former case the justification seems to be 'I have a certain disorder/disease', whereas in the latter it seems to be 'I do not want to the shortest person in my community'.

Even if some of the alternatives to the therapy-enhancement distinction mentioned above have contributed towards a better way of handling and understanding human enhancement, there is still much more work to be done in the area. Thus, we need to keep exploring different ways to frame and understand human enhancement, for instance one that promotes social values and aims to enable and empower individuals and society.

In this respect, the approach suggested here differs in at least two main aspects compared to other available approaches. On the one hand, it advocates a distinction not between enhancement interventions and non-enhancing interventions (whether these are seen as therapy, restorative or preventive), but rather a distinction between different kinds of enhancement interventions. The aim is not to keep building arguments on how to assess interventions in the grey therapy-enhancement area, but rather to suggest a different categorisation of human enhancement interventions. This approach acknowledges the difficulty in drawing a clear and acceptable distinction between enhancing and not enhancing, while at the same time recognising that the same intervention can enhance someone while making someone else worse off. On the other hand, the approach suggested here is different from previous ones in that it adds to the human enhancement discourse a view that is not based on individual, individualistic bodily based interventions to attain human enhancement. Instead it suggests a view in which human enhancement is based on the relationships shaping our lives. This suggested view about human enhancement takes on board the suggestions made by the discourse of ethics of care (Held, 2006; Slote, 2007) about the individual as a relational individual, immersed in complex and rich relations.

Furthermore, considering that the enhancement debate, which has been framed mostly under the therapy-enhancement distinction, has not fostered clear advances for policy-making, social action or a true commitment beyond the interests of a limited number of wealthy Western individuals, we can question the reasons for continuing to discuss ethical implications based on this understanding. Thus this book aims to move on and provide additional insight into the current enhancement debate by suggesting different understandings of human enhancement, in order to foster a more ethical way to use emergent technologies in the quest of improving the human condition.

Human enhancement entails difficult issues and questions that cannot be addressed in full within one single publication; as such, the main purpose of this book is to develop a coherent viewpoint and make a modest contribution to the ongoing discussion. The goals of this book are synthetic and interdisciplinary, as human enhancement brings forward pressing and important issues that cannot simply be profitably addressed from within the bounds of a single discipline. The suggestion of three different human enhancement paradigms is wholly constructive, as this area is clearly one where research is constantly been produced. Thus this book hopes to shed some new insights and keep the debate on human enhancement moving forward. The proposed typology is also imperfect, and should be viewed as a heuristic framework that makes progress and sets cautionary boundaries to the claims that can be made.

Secondly, the idea of different paradigms is used here to suggest different accepted points of view and modes of understanding. Each human enhancement paradigm suggested here reflects particular values and actions, and expresses different views on the legitimacy of problems and the proposed solutions to them. Thomas Kuhn (2012) in his book *The Structure of Scientific Revolutions* argued that new scientific breakthroughs create new paradigms or new perspectives linked to a set of practices. Thus, the main idea is that new technological and scientific breakthroughs, as those instantiated by emergent technologies, will bring or have already brought to the fore new paradigms. Within a new paradigm, old terms, concepts, understandings and experiments fall into new relationships one with the other (Kuhn, 2012). Here, it is argued that this is also true for our understanding of human enhancement.

The use of the term 'paradigm' here reflects some, but not all, of Kuhn's (2012) ideas. For instance, differences between paradigms are necessary but not necessarily irreconcilable. Therefore, it is expected that when changing from one paradigm to another, there will be a significant shift in the values determining the scope and goals of human enhancement and the legitimacy of proposed ways in which to enhance humans. However, in some cases the changes will be mostly only in scope and degree. In addition, if a human enhancement paradigm is to be accepted as the predominant one it should be regarded as 'more successful than its competitors' (Kuhn, 2012, p. 24), for instance by having better explanatory power or fitting better with our intuitions and values.

The main reason for putting forward three different human enhancement paradigms—the biomedical, the transhumanist and the social—is because these three different approaches cover, in an acceptable way, the different key ideas and values that can be found in the existing literature on human enhancement (in the case of the biomedical and the transhumanist) and the ones that have been lacking (in the case of the social). While some people might not fully agree with the terminology used here, it is important to keep in mind that it is only intended to be umbrella terms to cover sets of ideas, values and views. Thus, the suggestion here is not that these are the only possible human enhancement paradigms, rather that they can be seen as a reasonable starting point.

Thirdly, although there are many technologies that can be used for the purpose of human enhancement, the focus in this work is on emergent technologies. One reason for focusing on these technologies is that they are a clear example of a cluster of technologies that have gained sustained attention in the past years and have been suggested as tools in the quest for human enhancement. The scope and reach of these technologies have been perceived as having the potential to change the human condition in unprecedented ways. While the use of some of them has been debated in the past, such as pharmaceuticals in the area of brain enhancement, there are still many other enhancement interventions enabled by other technologies, in which ethical implications have not yet been widely discussed; for instance, those connected to methods of electrical stimulation of deep brain structures for enhancing mood and behaviours in healthy individuals, or the use of augmented environments supported by artificial intelligent systems.

One further reason is that emergent technologies have become an important player in the political discourse, garnering substantial amounts of federal and private resources. Finally, in societies where mental capital and well-being—understood here as an individual's 'cognitive and emotional resources' and 'the dynamic state in which the individual is able to develop their potential, work productively, and creatively' (Stewart, 2009)—have become extremely important for active participation in society, human enhancement technologies that promise to improve these areas will presumably be very attractive to many people. In addition to this, consideration of other widespread social tendencies, such as higher competitive pressure (even if it is only subjectively perceived) and the trend towards a 24/7 society, should give us enough grounds to carefully scrutinise interventions aimed at enhancing the human condition.

Finally, the scope of this project is unfortunately still largely limited to a Western perspective (though the book whenever possible has tried to include wider cultural perspectives of the issues involved). The book avoids making strong normative claims from any sort of consequentialism, Kantian ethics or Aristotelian ethics; but uses different relevant parts of these frameworks (a pluralistic approach) to assess the benefits and harms for the individual and society of the issues analysed. However, it does follow closely the normative framework suggested by recent work on the ethics of care (Held, 2001, 2006), in particular its view of the individual as relational in contrast with the liberal individual view. The latter is a view developed primarily for liberal political and economic theory, but when followed in other fields it has led to a seriously deficient concept of the individual. In addition, this book takes a pragmatic view in that it seeks to move the focus of human enhancement interventions towards areas that need to be enhanced but have been neglected.

The methodology not only exposes some of the main arguments for and against human enhancement, but also analyses them according to the different human enhancement paradigms. Thus, this work takes a contextualist approach to how different values shape different understandings of what is ethical and what is not when it comes to enhancement interventions, and about the different roles that emergent technologies can take within human enhancement. It discusses applications that are already available for certain groups of people (some of which may already be familiar to the reader), but also applications that are still being researched, only tested on mice or monkeys, or even just entertained in the minds of a few individuals. The discussion of emergent technologies that are or might be used for human enhancement is based on a number of overviews and systematic studies of the state of the art in human enhancement technologies as well as on other pertinent literature. The latter includes works with strong normative claims for and against human enhancement and relevant journalistic accounts.

Emergent technologies

Technology has been part of human history from the very beginning. Nonetheless, the way in which humankind approaches, perceives and uses technology has been dynamic throughout time. The growth and direction of technology is not only shaped by its interaction with science, but also by the market, society, law and politics. Thus, recognition of the social, economic and political dimensions of technology are crucial for understanding not only how we shape our technology, but more important how our technology shapes us (Bijker, Hughes, Pinch, & Douglas, 2012; MacKenzie & Wajcman, 1985; Sismondo, 1993).[5]

Another important feature of technology is *convergence*. Molecular biology and mechatronics are examples of the convergence of previously separate domains and disciplines. Goorden and colleagues have mentioned that there 'is a trend towards the merger of new sciences and technologies into what has been called "technosciences"' (Goorden, Oudheusden, Evers, & Deblonde, 2008, p. 215). These 'technosciences' are part of the convergence trend, and therefore are often referred to as 'converging technologies'. According to the High Level Expert Group (HLEG) 2004 report 'Converging Technologies: Shaping the Future of European Societies', convergence is used to refer to 'a merging of concepts from different systems of knowledge' (Nordmann, 2004), which can be shared devices and practices, the unification of previously separate domains of inquiry or a common goal that is approached from different directions. In the literature these converging technologies are also addressed as emergent technologies (Arnall, 2003; ETC Group, 2003; Kearnes, Macnaghten, & Wilsdon, 2006; Sarewitz & Karas, 2006;

Swierstra & Rip, 2007), enabling technologies and even disruptive technologies. While 'emerging' is used to connote that a certain technology is rising up or becoming important and prominent, 'disruptive' connotes an interruption, a disturbance to the orderly progression of an event, process or activity. Thus an emergent technology may become disruptive sometime, somewhere, even if its potential for such disruption may not have been recognised when it was first used (National Research Council, 2009).

In recent years, as part of a miniaturisation trend, the term 'converging technologies' has taken specific meaning through nanotechnology, which brings about a convergence of domains (Bainbridge & Roco, 2005; Canton, 2004; ETC Group, 2003; Nordmann, 2004; Roco & Bainbridge, 2002; Roco, Bainbridge, Tonn, & Whitesides, 2013), domains that work or aim to work at the nanoscale.[6] In other words, nanotechnology enables us to engineer and work at the level of atoms and molecules. So, for instance, what used to be separate domains, such as biomedicine, chemistry, physics, biology, materials science, electronics, information technology and optics, come together under the umbrella term 'nanotechnology' in a single engineering model.

The term 'converging technologies' did not only follow from the unification of different domains, but rather as the realisation that, besides nanotechnology, there were other enabling technologies and knowledge systems. By enabling technologies and knowledge systems, it is understood that they enable 'technological development on a broad front' (Nordmann, 2004, p. 13). Hence, as the HLEG Report explains, these technologies and knowledge systems are ready to enable one another and open up to new research and development challenges.

The meaning of converging technologies as a set of enabling technologies and scientific knowledge systems that converge in the pursuit of a common goal was established in the workshop *Converging Technologies for Improving Human Performance—Nanotechnology, Biotechnology, Information Technology and Cognitive Science (NBIC)*, organised by the US National Science Foundation and Department of Commerce in December 2001 (Roco & Bainbridge, 2002). In this report, the technologies mentioned are grouped under the 'NBIC' acronym, which stands for Nanoscience and nanotechnology (enabling other technologies by providing a common framework for all hardware-level engineering problems and contributing to the miniaturisation demands of other technologies); Biotechnology and biomedicine, including genetic engineering (enabling other technologies by identifying algorithmic structures and physical-chemical processes in living systems, and providing knowledge of cellular and genetic organisation); Information technology, including advanced computing and communications (enabling other technologies through their ability to model processes and represent physical states as information, and providing the computing power needed in all technical disciplines); and Cognitive science, including cognitive neuroscience (enabling the exploration and manipulation of the mind, and providing knowledge on how to acquire, represent and manipulate knowledge).

The use of the term 'converging technologies' from here on has followed a similar pattern, in that it describes a set of technologies that enable each other while working towards the same goal. Examples of this usage of convergence are found for instance in Douglas Mulhall's book *Our Molecular Future: How Nanotechnology, Robotics, Genetics and Artificial Intelligence Will Transform Our World* (2002). In this book Mulhall uses the acronym 'GRAIN', Genetics, Robotics, Artificial Intelligence and Nanotechnology, to cluster the set of technologies that he envisions will together transform our world. Another example is a 2003 Canadian science and technology foresight report which mentioned nanotechnology, ecological science, biotechnology, information technology and cognitive sciences as converging technologies for biohealth, eco and food system integrity and disease mitigation (Bouchard, 2003). The Action Group on Erosion, Technology and Concentration (ETC Group) in its 2003 report, 'From Genes to Atoms: The Big Down Atomtech, Technologies Converging at the Nano-scale', used the term 'Atomtech' to refer to the idea that converging technologies converge at the level of atoms and molecules. The ETC Group has also used the term 'BANG technologies', to convey the idea that information technology uses Bits, nanotechnology manipulates Atoms, cognitive technology manipulates Neurons and genetics manipulates Genes (2005).

In this book the term 'emergent technologies' will be used to cluster the type of technologies mentioned above, converging, emerging and disruptive technologies. In the following sections two of the key emergent technologies in the quest for human enhancement will be covered; some of the other emergent technologies will be briefly mentioned.

Nanotechnology

Nanotechnology is a nanoscale technology insofar as *nano* refers to the scale, in contrast to, for example, biotechnology, in which *bio* refers to life. The fact that nanotechnology is a technology concerned with the 'nanoscale' has allowed for its use as an umbrella term covering a heterogeneous set of technologies, sciences and engineering fields (Berne,

2006; Swierstra & Rip, 2007). It is partly owing to nanotechnology's interdisciplinarity and features as a converging technology that it has become such an important technology in the last decades.

The prefix *nano* derives from the Greek noun *nanos*, meaning 'dwarf' (Mader et al., 2006; Türk, Knowles, Wallbaum, & Kastenholz, 2006), or 'very short man' (Lin, 2007), and it is used to indicate that the measure corresponds to a one billionth part of the unity, for example, a nanometre (nm) is equal to one billionth of a metre. Thus, nanotechnology can be defined as a doctrine or theory of the artificial mastery of very small things. In nature the world of the very small is the world of atoms and molecules; therefore, nanotechnology can be understood as the set of techniques for manipulating atoms and molecules (Arnall, 2003; Bainbridge & Roco, 2005; Drexler, 1986; ETC Group, 2003, 2005; Nordmann, 2004; Roco & Bainbridge, 2002).[7]

To have an idea of what we are talking about, the thickness of a human hair is around 80,000 nm, the diameters of individual atoms range from 0.1 to 0.5 nm, and a DNA molecule is about 2.5 nm wide. While everything at the nanoscale is invisible to the human eye, in the last 100 years chemists, physicists and biologists have moved from debating the structure and composition of nature to working with nanoscale objects, as instruments to manipulate and to observe nature at that scale have become available (Tegart, 2006). The manipulation of atoms and molecules was first suggested in 1959 by the Nobel laureate Richard Feynman in his lecture 'There's Plenty of Room in the Bottom' to the American Physical Society at its meeting at Caltech in Pasadena, California:

> The principles of physics, as far as I can see, do not speak against the possibility of manoeuvring things atom by atom. It is not an attempt to violate any laws; it is something, in principle, that can be done; but in practice it has not been done because we are too big.
>
> (Feynman, 1960, p. 36)

However, the use of the term 'nanotechnology' as such is accredited to Norio Taniguchi of Tokyo Science University, who used it in his conference paper 'On the Basic Concept of "Nano-Technology"' (1974). Nonetheless, it was not until a few decades ago, with the invention of the scanning tunnelling microscope, Eric Drexler's controversial book *Engines of Creation* (1986), governments around the world spending millions in research funds for nanotechnology and the increasing number of different innovations using nanotechnology (from golf balls to self-cleaning paintings), that 'nanotechnology' became a commonly used term and the subject of large debate. Nowadays there are an increasing number of specialised institutes and university programmes that focus on nanotechnology research and development, and an increasing number of special reports, courses, conferences and academic journals addressing nanotechnology. There are even science fiction novels, such as *Prey* by Michael Crichton, and movies, such as *Spiderman* and *Minority Report*, that have mentioned nanotechnology.

Definitions of nanotechnology vary depending on national strengths and lines of research. Some definitions make a distinction between nanoscience and nanotechnology (Royal Society, 2004), others draw attention to the novel properties and functions it enables (NSTC, 2002). The International Organization for Standardization (ISO) definition for the standardisation of nanotechnology covers either or both of the following:

- Understanding and control of matter and processes at the nanoscale, typically, but not exclusively, below 100 nanometres in one or more dimensions where the onset of size-dependent phenomena usually enables novel applications,
- Utilising the properties of nanoscale materials that differ from the properties of individual atoms, molecules, and bulk matter, to create improved materials, devices, and systems that exploit these new properties.[8]

The ISO's nanotechnology definition has the potential to influence greatly the global course of development in the field, as it allows for a more international and collaborative understanding of the term and its scope. However, some scholars, such as Joachim Schummer, have argued that the lack of a meaningful nanotechnology definition has allowed for a large amount of cutting-edge research to be relabelled as 'nano' in almost all science and engineering disciplines, 'without having much new in common and without showing any remarkable degree of interdisciplinarity' (Schummer, 2004).[9] This relabelling can indeed be understood not only as the result of a lack of a meaningful definition, but also as a way to obtain funding resources for particular research projects, whether they are or are not definitively connected to nanotechnology (Wolbring, 2008b; Wolpe, 2002).

As an emergent technology, nanotechnology plays a special role owing to the miniaturisation trend and its capacity to enable us to manipulate matter at the nanoscale. Connected to the former, one of Kurzweil's predictions is that almost all technology will be nanotechnology by the 2020s, because 'both electronic and mechanical technologies are shrinking at a rate of 5.6 per linear dimension per decade' (Kurzweil, 2007, p. 41). Connected to this, it can be said that a great deal of nanotechnology's unique power and potential comes from the fact that at the nanoscale the properties of elements and materials can change dramatically from those they feature on a larger scale. That is to say, with nanotechnology size does matter. Many of the properties of a particular substance that we take for granted at the micro and macro scale, with only a reduction in size can change completely, showing different electrical, conductivity, colour, strength, elasticity, melting point and reactivity properties. For instance, carbon in the form of graphite is soft and malleable whereas at the nanoscale it can be stronger than steel. Nanoscale particles of aluminium oxide are used as explosives, whereas on a larger scale dentists use them to repair teeth. Many of these changes seen at the nanoscale are the result of 'quantum effects' (Allhoff, Lin, & Moore, 2009; Ratner & Ratner, 2003; Wilson, Kannangara, Smith, Simmons, & Raguse, 2002).

Another reason for considering nanotechnology as a special technology is the precision and specification that nanotechnology puts into human hands, which allows us to visualise, measure, monitor, simulate, manipulate, comprehend and improve matter (Allhoff et al., 2009; Allhoff, Lin, Moor, & Weckert, 2007; Berne, 2006; Drexler, 1986; ETC Group, 2003, 2005; Freitas, 2005a; Kearnes et al., 2006; UNESCO, 2006; Wood, Jones, & Geldart, 2003).

This view is exemplified in the following statement:

> The emerging fields of nanoscience and nanoengineering—the ability to precisely move matter—are leading to unprecedented understanding and control over the fundamental building blocks of all physical things.
>
> (Roco, Williams, & Alivisatos, 2000)

That is why molecular nanotechnology—which approaches the way in which biological systems work—has been seen as an approach that could offer us unique ways to substitute equivalent alternatives and even to create organic and inorganic structures, molecule by molecule or even atom by atom, following nature's way (Amato, 1999; Arnall, 2003; Drexler, 1986). As Drexler put it, nanomachines or nanobots will let us build almost anything that the laws of nature allow to exist or that we can design, as we 'will be able to bond atoms together in virtually any stable pattern, adding a few at a time to the surface of a workpiece until a complex structure is complete' (1986). According to Drexler, there is nothing fictional about this idea given that nature already works in this way. However, Richard E. Smalley, an American chemist and physicist awarded a Nobel Prize for the discovery of buckyballs,[10] did not quite agree with Drexler on this. According to Smalley, 'self-replicating, mechanical nanobots are simply not possible in our world' (2001, p. 77) because of what he calls the 'fat and sticky fingers' problems, which according to him are both fundamental and unavoidable. The 'fat fingers' problem has to do with there not being 'enough room in the nanometre-size reaction region to accommodate all the fingers of all the manipulators necessary to have complete control of the chemistry'. The 'sticky fingers' problem, according to Smalley, is that 'the atoms of the manipulator hands will adhere to the atom that is being moved. So it will often be impossible to release this minuscule building block in precisely the right spot' (2001, p. 77). Through experimentation it was later proven that both of these problems are not fundamental and are avoidable (Drexler, Forrest, Freitas, Hall, & Jacobson, 2009; Lee & Ho, 1999). This implies that nanotechnology has indeed the potential to be applied to virtually any manufactured goods across all industry sectors, bringing 'new platforms for industrial manufacturing that could make geography, raw materials, as well as labour, irrelevant' (ETC Group, 2005, p. 4). Hence, nanotechnology could bring not only quality changes, as those brought up by the quantum effects, but also bring quantity changes, as 'bottom-up' self-assembly develops.[11] However, even though self-assembly—or molecular nanotechnology as it has also been labelled—is theoretically possible, we have not yet reached self-assembly for the manufacturing of complex devices and materials. As the Triennial Review of the National Nanotechnology Initiative states, 'there is still a gulf between this vision and popular images of nanotechnology in which the bottom-up approach is routinely used to manufacture complex, large-scale industrial objects such as computers or buildings at very low cost' (CRNNI, 2006).

In addition, there are still numerous technical challenges ahead of us. For instance, despite the fact that nanotechnology can increase the amount of electronics in ever-smaller devices, there might be a limit to how small things can get without losing quality or functionality. Connected to this point, a group of researchers at Umeå University in Sweden and the University of Maryland in the US have claimed that when the size of components approaches the nanometre scale, information will disappear before it has time to be transferred (Marklund,

Brodin, Stenflo, & Liu, 2008). Other technical difficulties to be overcome are the tendency of nanoparticles to aggregate either under physiological conditions or during storage (Tomellini, Faure, & Panzer, 2006), or issues connected to power management (Nicolelis, 2003).

These challenges have not stopped a vast amount of nanotechnology investment, and research and development (R&D). The US National Science Foundation predicts that nanotechnology will capture a $1 trillion market by 2015 (Roco & Bainbridge, 2002). Investment and research areas include aerospace and defence, electronics and information technology and communication (ITC), chemical industry, energy, life sciences and healthcare, construction, textiles, environment and water, food and drink (production, processing and safety, packaging), consumer goods and household care, and automotive and transportation, as well as the wide variety of already available products on the market, including sunscreens, processors for electronic devices, waterproof and self-washable cloths, beauty and health creams, solar panels, drugs, lotions, sports equipment, adhesives, paintings and coatings, among many others.[12] Nanotechnology's R&D programmes have also become popular and highly integrative, bringing companies, both large and small, into research partnership with universities, they involve multiple funding agencies (government, industry and international organisations), have a long-term planning perspective, have led to an increasing number of specialised organisation devoted to technology transfer (with various degrees of government involvement), and in some cases to agencies or programmes supporting the societal and ethical dimension studies too (Roco & Bainbridge, 2002). All these features of nanotechnology R&D and investment have made nanotechnology a more interesting and controversial field than biotechnology and a strategic platform for global control of food, health, manufacturing, energy, agriculture, computers and electronics. This is reflected in the increasing number of nano-related patents (ETC Group, 2003).

Hype or not, it still remains true that nanotechnology has great potential to change the human condition. Firstly it allows us to exploit the unique properties that occur only at the nanoscale and it allows us to manipulate and control atoms and molecules. It offers a broad technology platform for industry and holds promises for applications that have the potential to manage many economic, technical, ecological and social problems. It has also stimulated many countries and companies to invest in the field, has changed and broadened manufacturing capabilities, and has become one of the main drivers for economic, technological, cultural and political change.

Neurotechnology

Neurotechnology emerged as a result of current knowledge and techniques in modern neuroscience and a combination of traditional approaches (Banks, 1998; Bear, Connors, & Paradiso, 2007). Current advances in nanotechnology and other emergent technologies have enabled the rapid grow of neurotechnology. For example, neurotechnology helped by nanotechnology can enable us to gain insight into the different molecules and proteins involved in brain function (Lynch, 2004), as well as enable us to monitor changes in the human brain with improved spatial and temporal span (Farah & Wolpe, 2004; Kandel, 2000).

Remarkably, many of the key structures of the vast human nervous system exist at the nanoscale, such as the vesicles that store neurotransmitters, the gap between neurons across which those neurotransmitters flow and the pigment molecules in the eye that make vision possible (Bainbridge & Roco, 2005, p. 2).

Neurotechnology is the most rapidly advancing area of medicine and biotechnology (Coenen et al., 2009; Glannon, 2006). Neurotechnology has already given us significant insights into areas such as genetics (identification of disease genes that play an important role in several neurodegenerative disorders), gene-environment interactions (environmental influences such as toxic substances, diet and level of stress can trigger certain traits of an individual), brain plasticity (ability of the brain to modify its neural connections), new drugs (new insights into the mechanisms of molecular neuropharmacology provide new treatments), brain imaging techniques, cell death (insights into how and why neurons die) and brain development (Carey, 2008; Kandel, 2000).

'Neurotechnology' is an umbrella term used to describe a wide range of technologies, such as neuropharmacology, Magnetoencephalography (MEG), functional Magnetic Resonance Imaging (fMRI), brain-machine interfaces and cognitive and sensory prostheses. In contrast with nanotechnology, in which *nano* is concerned with the scale, in neurotechnology, *neuro* is used to make reference to the study and manipulation of the nervous system, including the brain. Definitions of neurotechnology vary, but unlike the ISO's standard definition for nanotechnology, no standard definition for neurotechnology yet exists. However, most definitions agree that neurotechnology is technology that deals with the human nervous system. For example, the US National Neurotechnology Initiative (NNTI) Act defines neurotechnology as both the science and technology that allows an individual to understand, analyse, heal and treat the brain and nervous system (Neurotechnology Industry Organisation, n.d.).[13] Neurotechnology's main fields of research are (1) neuropharmacology,[14] (2) neurodiagnostics (which includes neuroimaging, in vitro diagnostics and neuroinformatics) and (3) neurodevices (which includes neuromodulation, neural interfaces, neurofeedback and neurosurgical devices).[15] And its applications go beyond just therapeutical ones, to military, entertainment (e.g. video games), neuroeconomics, neuromarketing and education.

Although neurotechnology was not directly mentioned in the earlier reports on converging technologies and emergent technologies—such as the Roco and Bainbridge edited reports or any of the ETC Group reports—it has started to become mainstream and to appear more often in policy reports and journal articles (Bioethics, 2013; National Research Council, 2008; PCSBI, 2014; Roco et al., 2013). A recent expert meeting held in Brussels entitled 'Shifting Boundaries, Changing Concepts: The Challenges of Human Enhancement to Social, (Dis-)Ability, Medical and Ethical Frameworks' showed that there is wide consensus from the experts regarding the idea that neurotechnology clearly has the potential for 'a new quality of interventions into the human mind and body' (Coenen et al., 2009, p. 159).[16] Some people even regard neurotechnology as the 'next wave of techno-change' (Lynch, 2004).

One of the main reasons why neurotechnology is considered key in the human enhancement quest is its subject of manipulation, namely the human brain and nervous system. While we generally consider our genes special because they are part of us, we tend to consider our brains as more special given that we regard them as part of our essence, that is to say as having a direct connection with our sense of self and identity, going beyond the biological. To illustrate this point, consider the remark of Donald Kennedy, former Science magazine's editor-in-chief: 'I already don't want my employer or my insurance company to know my genome. As to my brainome, I don't want anyone to know it for any purpose' (Kennedy, 2003). Furthermore, the brain is considered to be the bodily 'centre of control', the organ in charge of affective and cognitive capacities of reasoning and decision-making (Berger, Gevers, Siep, & Weltring, 2008; Carey, 2008), the most dynamic and sensitive part of our anatomy and physiology to intervention, and the most important human organ (Carey, 2008; Hippocrates, n.d.). Thus, intervening directly in the brain is regarded as having more far-reaching consequences for human interaction (Berger et al., 2008) and for human self-perception and understanding than any other intervention in our body (Farah, 2010; Glannon, 2006; Racine, 2010; Wolpe, 2002). Neurotechnology can help us to understand and alter the brain with detail and depth that previous technologies have not enabled us to achieve before; for example, it can enable us to control in new ways external devices as well as bring forward new modes of perception (Coenen et al., 2009). Neurotechnology also has the potential to shed light on our long-standing questions about one of the most complex organs and about the different features we ascribe to it (intelligence, identity and the like).

Articles and reports regarding neurotechnology are in the increase as well as neuro-related patents (National Research Council, 2008). Some people think we are seeing the rise of a neurosociety (Lynch, 2004), in which understanding of our identities and our 'selves' is being described increasingly in terms of 'neuro'. We have become 'neuro selves', as Nikolas Rose puts it:

> [...] a neurochemical sense of ourselves is increasingly being layered onto other older senses of the self, and invoked in particular settings and encounters with significant consequences.
>
> (2007, p. 222)

Summing up, it seems reasonable to say that neurotechnology is indeed a ground-breaking technology, firstly because there is a rapidly growing market for its products, driven by global public demand and international market forces such as more people living longer, a market on life-style improvements, and the worldwide economic burden of neuro-related conditions (National Research Council, 2008). Perhaps more importantly, it is also because its subject of study and manipulation is the human brain. Thus its importance emerges not necessarily because it will directly change our world and environment, as in the case of other technologies; but because it has the potential to change the way we see and understand our world, environment and ourselves by changing us from the core.

Information Technology

Information technologies are technologies that generate, transmit, receive, store, process, access or analyse data. The developments in mathematics, manufacturing, materials sciences, media and many other areas have helped the advancement of information technologies, and the economic impact of improved computing and communication has accelerated it (Roco & Bainbridge, 2002). Information technology enables other technologies through their ability to model, process and represent information (Roco & Bainbridge, 2002). As such these technologies are not about the scale or the specific object of study and manipulation, but about the common language of information and communication, namely the bit-based language (at least that has been the predominant one until recently).

These types of technologies have had a large impact on society, from writing to the printing press, and now to computer- and digital-based information technologies. The Internet, smartphones, tablets, augmented reality, for example, have transformed how people work and interact with each other; we can even say that these routinely support various human activities far outstripping our biological cognitive capabilities. These technologies, and their increasing power to collect, store, share, visualise and analyse data have accelerated research and development in other areas, from neuroscience to robotics (Sandler, 2014), and to date are considered to be the technologies that have produced the most dramatic advances in terms of the ability to process information (Bostrom and Sandberg, 2009a). For example, data mining and information visualisation help us to process and display enormous amount of information in ways that are easier to understand, common software tools help us to keep multiple items in our memory to use later, support our decision-making capabilities and perform other routine tasks (Bostrom and Sandberg, 2009a; Roco & Bainbridge, 2002).

One main trend in current information technologies is their integration level with the user. An example of this is ubiquitous computing or 'ubicomp' (also referred to as smart computing, ambient intelligence and the internet of things), a computing paradigm in which the computer has disappeared in the background, and ubiquitous, invisible machines are embedded in environments and everyday objects such as pens, books, watches, buildings, walls, furniture and clothes, serving and actively supporting the user (Kuniavsky, 2010). Thus, the computationally enhanced future that ubiquitous computing promises is not so much focused on machines (intelligent or otherwise), but on people themselves.

The ubiquity of smartphones in every aspect of our lives could be seen as evidence that we are already immersed in a computational experience that blends the virtual and the real. Other examples are devices such as smart eyewear, which promises to bring information in front of our eyes, or smart watches which track the wearer's movements and monitor their health and fitness. Industrial and academic centres have been focused on developing smart living spaces that can monitor our everyday activities in order to provide improve comfort, energy effi-ciency and security or monitor the health and general well-being of senior citizens, as well as smart environments that monitor air and water quality, atmospheric conditions and wildlife populations. The leverage that this new paradigm is garnering can be seen from the research clus-ter and initiatives that it has triggered, such as the European Research Cluster on the Internet of Things (IERC), which was established back in 2007.

While, the increase in information and communication technologies has had tremendous social impacts, such as helping to connect the global community, it also has helped to erode the values of small com-munities, and brought more threats to privacy. It has brought to the fore issues about surveillance, security, trust and digital divides—the division in the world between those who have access to new information and communications technology and those who do not (Quibria, Ahmed, Tschang, & Reyes-Macasaquit, 2002).

Biotechnology

The prefix *bio* in biotechnology makes reference to life, thus biotechnology deals with the use of living organisms, biological com-ponents and biological processes to create useful products. Here, as in previous cases, the term 'biotechnology' covers a broad range of activ-ities that are related through the fact that they involve the creation, use or manipulation of 'biological' components, biological processes or systems (Bioethics, 2012), including genetic and genomics tech-nologies, synthetic biology, tissue engineering, genetic engineering, pharmaceutical biotechnology, reproductive technologies and stem cell technologies.

The applications of biotechnology can be found in almost every field of human activity that is important to our well-being and way of life, including medicine, food and energy production and industry. Work has targeted the creation of 'bio-nano processors' for programming complex biological pathways that can mimic cellular processes on a chip, as well as ongoing work to understand how genes are expressed in the living body (Roco & Bainbridge, 2002).

Biotechnology has grown owing to the confluence of different fields, including biology, advanced computing, pharmacy and others, and has been accelerated by its expected benefits and political investment placed in it. For example, biotechnology is held up as an impor-tant source of future remedies for current challenges and crisis, from fuel and food security to personalised medicine and environmentally sustainable growth. However, just as with the other emergent technologies mentioned here, in spite of the substantial benefits from biotechnologies they also come with negative impacts for society and individuals. The development and use of biotechnologies gives rise to several morally relevant questions, such as the distinctive significance that is attached to living things and the limitations of human understanding or control over biology. Other concerns that become salient within biotechnologies are 'biosafety' and 'biosecurity', as well as 'dual use', that is to say when biotechnologies are used for malign as well as benign purposes (Bioethics, 2012).

Robotics and artificial intelligence

Robotics is a branch of technology and engineering involving the conception, design, manufacture and operation of robots. Broadly speaking, robots are machines endowed with sensing, information processing (including perception, reasoning, planning, learning, feedback signal processing and control), and motor abilities that enable them to achieve goal-oriented and adaptive behaviours (Tamburrini, 2009). New generations of robots are becoming increasingly capable in coordinating behaviours with heterogeneous teams of agents that include other robots, humans and software systems. Advances in other fields such as electronics, computer science, artificial intelligence, mechatronics, nanotechnology and bioengineering have enabled advances and new developments in robotics. For example, communication technologies have enabled robots to access networks of software agents hosted by other robotic systems.

Just a few decades ago, robots were mostly confined to industrial environments; today, there are several ways in which humans use and interact with robots. For example, robots are used for border surveillance, handling of dangerous material, rescue missions, and for the exploration of environments that are hostile for humans (such as deep ocean and outer space); but also in more direct interaction with humans as they have several applications at homes (doing household chores), offices, hospitals, museums and schools, including education, training, healthcare and entertainment.

Artificial intelligence (AI), on the other hand, deals with the study, the development and the use of intelligent machines and software, which can help us better understand how to increase human's reasoning, learning, and perceptual processes. Highly intelligent AI, such as the one portrayed in science fiction movies, is still very much a futuristic vision shared by few commentators and does not yet depict reality. Nonetheless, what has been termed weak or narrow AI is already at the core of many aspects of our everyday life. For instance, different applications in the Internet, search engines and in smartphones rely on various forms of AI. Medicine, engineering, finance and entertainment as well as many military applications rely on AI. Current forms of AI can schedule appointments, allocate resources for large corporations, make financial and meteorological predictions, play chess or land aircrafts.

Thus, while we are not quite in a world in which fully autonomous robots and conscious, self-aware artificial intelligence are seamlessly integrated into every aspect of our lives, robots and intelligent machines have become commonplace in our lives. While concerns about the development and widespread dissemination of robots and AI have tended to focus on the risks involved, such as the scenarios portrayed in movies such as *Terminator*, or *I Robot*, where these malfunction and turn against us, or on the ways in which they could change our way of life for the worse, there are other more day-to-day ethical issues that arise as we built more intelligent and autonomous robots. Roboethics is the field of applied ethics that has set itself to analyse these ethical issues. There are also research networks, such as the European Robotics Research Network (EURON), which have incorporated plans to develop guidelines for programming and designing robots—including areas such as safety, security, privacy, identifiability and traceability—so as to avoid detrimental outcomes. Other key issues that robotics and AI bring to the fore are related to moral responsibility, liability and ethical dimensions dealing with personhood and agentivity (Tamburrini, 2009).

What makes emergent technologies important

Among the features that make emergent technologies important we can mention the following:[17]

– Embeddedness: In a context in which emergent technology-based applications surround us more and more without us noticing their presence and our dependence on them, the more we can say they are forming 'an invisible technical infrastructure for human action' (Nordmann, 2004, p. 3). Given the level of dependence and infiltration of these technologies in our lives, it is plausible to agree with Neil Postman, who once said that 'technological change is not additive; it is ecological' (1998, n.p.), meaning that its consequences do not only add something but rather that they change everything.
– Unlimited reach: As these technologies enable each other it appears that we can engineer everything, from our own bodies and minds to our social interactions and our environments.

- Specificity: These technologies also have the potential to enable us to address very specific tasks, such as combating cancer cells without having to kill healthy cells.

These features are regarded as important within the discourse of human enhancement because they can help in areas such as improving learning and work efficiency, enhancing individual sensory and cognitive abilities, improving group and individual creativity and communication, and ameliorating or preventing physical and cognitive decline. All these payoffs can be summarised within the main goal of improving the human condition (Bainbridge & Roco, 2005; ETC Group, 2003, 2005; Miller & Wilsdon, 2006; Roco & Bainbridge, 2002; Roco et al., 2013; Sarewitz & Karas, 2006).

Emergent technologies are also important because of the different understandings they instantiate, for instance as revolutionary technologies, or as socially robust in as much as these are constructed by public interaction with the development of emergent technologies (c.f. Kearnes et al., 2006). Connected to the idea of revolutionary technologies, Kevin Kelly nicely captured the idea that if technologies do not create worries for us, then they are not revolutionary enough (2006).

These technologies have also been regarded as science fiction and cyborg/posthuman technologies (Cabrera, 2009a; Milburn, 2002); as science fiction because of the trend of speculation and prognosis that is noticeable in a large number of writings and public understandings regarding these technologies (Arnall, 2003; Drexler, 1986; Freitas, 2007; Garreau, 2005; Gordijn, 2005; Schneider, 2010; Wood et al., 2003). This of course has brought to the fore many difficulties regarding decisions on which projects should be funded or not (Nordmann, 2007), as well as sparking comments about the need to separate between science and science fiction applications, in particular to move forward with real practical advances (Lin & Allhoff, 2006).

Donna Haraway (1991) and Scott Bukatman (1993) have suggested that science fiction technologies transfigure embodied experience, creating the appearance of a posthuman subject. Thus, these technologies have also been understood as cyborg/posthuman technologies, in particular because of their potential for blurring limits, not only between science fact and science fiction, but more importantly ontological limits such as the limits between us as humans and our technology (Cabrera, 2009a; Milburn, 2002).

These two particular understandings around emergent technologies tend to be either too pessimistic or too optimistic, or to overestimate certain facts while underestimating others (Allhoff, 2007; Blackman, 2009). Nonetheless, they have become part of its meaning, shaping the different roles people envision these technologies will play in human enhancement. All of this affects the way we balance the ratio between benefits, costs and risks; the way the issues are framed; and the ways we assess and discuss the challenges that these technologies bring forward. Thus, having a better understanding of the public's perceptions can illuminate the 'societal and ethical implications' of these technologies (Schummer, 2004, p. 56), providing us with a platform to explore our intuitions related to identity, dignity, autonomy and similar ethical salient concepts (Allhoff, 2007; O'Mathuna, 2009). Moreover, as Rebecca Roache has pointed out, arguing against those who think that we should not spend our energies and ethical attention in fictional and improbable future scenarios and speculations, without this exploration 'many of our most important ethical projects would never even have been entertained' (2008, p. 323). Hence, attending to possible future scenarios has the potential to help us address current issues we are faced with today and to develop the moral tools to do so. This can be done without neglecting the importance of separating real potentials from imponderable possibilities.

All these different features and understandings of emergent technologies make them groundbreaking technologies, in the sense that they are reconstructing our world and our understandings of it and ourselves.

Ethics of emergent technologies

The aforementioned characteristics of emergent technologies increase its potential benefits as well as its dangers, producing new or exacerbating previous ethical and social challenges. Let us not forget that progress is not only about great technology (Hurd, 2005); 'with great power there must also come great responsibility'.[18] This last point is of particular importance when considering the possibility that the scope and speed of research and development around these technologies might exceed regulators', social scientists', scientists' and ethicists' capacities to assess their impact.

Ethical issues concerning these technologies have instantiated separate fields of ethical inquiry, such as nanoethics (Allhoff, 2007; Moor & Weckert, 2004; O'Mathuna, 2009), neuroethics (Farah, 2010; Glannon, 2006; Illes, 2006; Levy, 2007; Racine, 2010; Roskies, 2010) and roboethics (Lin, Abney, & Bekey, 2012; Sawyer, 2007; Tamburrini, 2009). While there is debate about whether or not these technologies should have separate branches of applied ethics, as it is not clear that the issues will be new in any interesting sense (Allhoff, 2009b; Holm, 2013; Keiper, 2007; Safire, 2002; Wolpe, 2010), we can still agree with the idea that given their particular features they can exacerbate and make more urgent past issues (Lin, 2007; Weckert, 2007). For some, the main ethical challenges posed by emergent technologies in the context of human enhancement are not only different in scope, number and depth, but more importantly in the very form that ethical reflection takes (Coenen et al., 2009; Farah, 2010; Fukuyama, 2002; Habermas, 2003; Roco & Bainbridge, 2007; Sandler, 2014). Thus, whether or not different applied ethics fields are needed to study the ethical and social impacts of these technologies do not make the concerns less important. Yet, it is important to ensure that by having these distinct fields we are not 'reinventing the wheel' over and over again (Alpert, 2008); rather, the focus should be on how they can learn from each other and even work together to advance ethical analysis and debate. Particularly, with issues such as human enhancement, where competing and disputable interests are involved, a dynamic process of negotiation and exploration of values, meanings, purposes, intentions, directions and goals is needed. In this sense ethical enquiry never stops (Moor & Weckert, 2004); it is needed during the early stages of development as well as in dealing with the consequences of a given technology.

Conclusion

This first chapter has introduced and set the context and background for the topic this book deals with, namely the ethical considerations for choosing one kind of enhancement over another when using certain technologies. It has set the stage for the three human enhancement paradigms to be explored in this book: the biomedical, the transhumanist and the social. The aims and scope of the book were also stated here, in particular the aim to move the discourse on human enhancement from individualistic and high-technology applications to more inclusive and sustainable ones. It has concluded with an overview of some emergent technologies that are perceived to be key in the human enhancement endeavour, highlighting their importance, the significant amount of research and investment being undertaken and why ethical discussion around the issues they instantiate is needed and important, particularly with respect to human enhancement.

2
The Biomedical Paradigm

Health, disease and the goals of medicine

Of the three human enhancement paradigms to be discussed in this book, this chapter deals with the biomedical one. This paradigm is based on the therapy-enhancement distinction and is by far the dominating human enhancement paradigm within the enhancement discourse. In order to better grasp the values and perspectives held by this human enhancement paradigm, this chapter starts with an overview of the different concepts used within the biomedical enhancement paradigm, namely disease, health and the goals of medicine. These concepts are rarely included in ethical analyses of emergent technologies when used for biomedical or human enhancement applications. However, as Marianne Boenink and others have argued, these concepts are relevant to ethical debates on emergent technologies (Boenink, 2009; McKenny, 1997; Sandler, 2009). In particular, there are two main reasons for holding such a claim:

1. A better understanding of the concepts underlying the therapy-enhancement debate could help us determine whether or not we should keep debating issues around human enhancement based on these concepts.
2. Understanding the different concepts underlying the biomedical paradigm could enable a more fruitful debate regarding the desirability of the values promoted by this paradigm.

Before continuing, it is worth clarifying that the term 'biomedical paradigm' is used following a medical anthropology theoretical approach. Under such an approach medical practices are based on an ontological and epistemological dualism. The former view distinguishes the physical from the mental (body/mind distinction); whereas the latter is a newer kind of dualism referring to the distinction between disease and illness (Sullivan, 1986).[19] Both views are considered to be the basis of the Western scientific paradigm of modern medicine.

The concepts of health, disease and the goals of biomedicine are inter-related concepts. Diseases are generally related to unhealthy conditions or undesired conditions, the goals of medicine to the promotion of health and fighting disease. This section aims to shed light on why these concepts are not really suitable for evaluating the ethical issues brought forward by human enhancement.

Understanding disease

Under this paradigm, disease is described as the *objective* assessment of abnormal or pathophysiological phenomena,[20] displayed by an individual or group of living organisms, in connection with a specific common feature or set of features, by which they differ from the *statistically* defined norm of their species, placing them at a biological disadvantage (Boorse, 1977; Campbell, Scadding, & Roberts, 1979; Daniels, 2000, 2008). In the enhancement literature the terms 'ill health', 'illness', 'sickness' and 'disease' are generally used interchangeably or understood as similar terms describing deviation from health. However, strictly speaking these terms do not connote the same thing and thus should not be used interchangeably (Boorse, 1975; Clouser, Culver, & Gert, 1997; Eisenberg, 1977); more so in the context of human enhancement, where some of its definition problems arise precisely from considering different aspects of ill health.

Sickness, as used by medical anthropologists and some social scientists, refers to the public or social component of ill health, and includes both disease and illness (Thomas, 2003); illness is seen as the shareable experience of *subjective*—individual and mostly (but not necessarily) private—suffering and diminished capacity, and disease as the *objective* assessment of ill health. The distinction helps to explain why one can feel ill in the absence of disease or that one can have a disease and not feel ill (Eisenberg, 1977). Accordingly, illness cannot be diagnosed nor treated, while disease calls for treatment.[21] One of the main problems with overlooking these distinctions is the diagnosis and development of treatments for the non-normative component of sickness (Kirmayer, 1988; Scheper-Hughes & Lock, 1986, 1987). This means that within this paradigm ill health tends to be individualised and medicalised (Illich, 1975; Wolbring, 2005), rather than collectivised and politicised (Scheper-Hughes & Lock, 1986); transforming the social into the biological, and the biological into the technological. The problem with this view is that we run the risk of turning everything into a disease, which relates to problems of medicalisation (also known as disease mongering). While medicalisation is not inherently a bad thing (Parens, 2013), it is associated with the promotion of disease and illness by strategically exploiting and abusing people's anxieties, fears and prejudices in order to create markets for treatment (Conrad, 1992; Moynihan, Heath, & Henry, 2002; Nichter, 2008; Wolbring, 2005). Examples of medicalisation are reflected, for example, in the increasing number of people using Viagra and viewing certain natural processes of the body as diseases (such as menopause or ageing). Medicalisation is also connected with the creation of non-desirable views towards certain human conditions. However, it undermines the fact that some of the conditions that it labels as non-desirable might actually be needed or desirable under other circumstances. Consider the case of sickle cell trait, which causes debilitating symptoms at elevated altitudes, but can protect against malaria in lowland climates (Gallup & Sachs, 2001). At a deeper level, medicalisation 'may help to feed unhealthy obsessions with health, obscure or mystify sociological or political explanations for health problems, and focus undue attention on pharmacological, individualised, or privatised solutions' (Moynihan et al., 2002, p. 886).

Even if we leave aside the problems brought forward by overlooking the distinction between ill health and disease, there are further problems with the current biomedical perspective on disease. First, given that the biomedical model is based on a dualistic view of the body and the mind, it is assumed that the body can be understood and treated in isolation from other aspects of the individual in question (Scheper-Hughes & Lock, 1986; Thomas, 2003); second, it undermines the fact that many deviations from the average are not unhealthy, for example, unusual strength; and finally, its statistical basis makes it difficult to regard an entire population as having a disease.

Technology in biomedicine has brought forward even further challenges. For example, new technologies are used for measuring and assessing the markers and end points of disease. This has influenced explanatory models and shapes how we act towards disease (Hofmann, 2001). Thus, it is likely that new medical applications and developments in emergent technologies will reshape our concept of disease. For instance, Robert A. Freitas, a very active researcher and promoter of nanomedicine, argues that nanomedicine requires a new concept of disease that he refers to as the volitional normative concept of disease (Freitas, 1999, 2007). This concept of disease starts with biomedicine's definition of disease, but goes beyond it by including the patient's desires. It is out of the scope of this book to explain further Freitas's suggestion; however, it is worth pointing out that our understanding of disease has changed and will likely continue to change according to different sociocultural and technological tendencies.

Understanding health

Another concept that is crucial in the therapy-enhancement distinction is the concept of health. There are many different understandings of health. Some view it as the absence of pain, suffering or discomfort, others as a form of adaptation and some others as homeostasis. Some people see health as a normative concept (Agich, 1983; Nordenfelt, 2003), while others as a value-free naturalistic one (Boorse, 1977; Daniels, 1985, 1992). A general feature to all the different understandings of health is that it is regarded as valuable and as a desirable good that contributes to opportunity and social progress (CSDH, 2008; Daniels, 2008; WHO, 1946).

The World Health Organization's (WHO) definition of health is the most used in biomedicine.

> *Health is a state of complete physical, mental, and social well-being and not merely the absence of disease or infirmity.* The enjoyment of the highest attainable standard of health is one of the fundamental rights of every human being without distinction of race, religion, political belief, economic or social condition.
>
> (WHO, 1946)

This definition of health is a positive one, because it rests on the idea that health should be more than the absence of disease, combining both biological and social aspects of health. While acknowledging the mental and social dimensions of health is an important contribution to the common understanding of health (Bircher, 2005), such a positive view is not without challenges. First, it promotes an unlimited ideal, with no fixed goal of perfect health or a unique direction of advancement. This makes both health and complete absence of pathology practically unattainable (Illich, 1975), but it also contributes in blurring the distinction (if any) between therapy and enhancement. Second, it leaves us with no clear directions on how to balance health needs against available medical resources (Bircher, 2005), or on how to make decisions about which goals are worth pursuing when there are more than two incompatible goals to choose from.

Another problem with the current views on health is that they are associated with ideas about what is *normal* for humans; thus a common goal of modern medicine has been the restoration of *species-typical* functioning (Boorse, 1977; Daniels, 2008). However, basing our understanding of health on what is normal fails as a sufficient (unhealthy conditions may be typical within a society) or necessary (unusual conditions may be perfectly healthy, such as being red haired) condition of health. Moreover, the meanings of concepts such as 'normal' or 'natural' in our current society are increasingly open to debate. Consider the case of someone with an impairment. While that individual probably does not conform to the general view about what is 'normal', it does not follow that his or her condition is necessarily unnatural or that he or she is not a healthy individual. Furthermore, consider the case of individuals who are near the lower end of a normal health distribution curve. They might be considered to be at a disadvantage compared to those within the average of the distribution curve; but those individuals within the average could equally regard themselves at a disadvantage compared to those outliers above the average (PCB, 2003). Subsequently, as we strive to bring those outliers in the lower end closer to the average, the normal distribution curve is shifted, and with it what we regard as normal. Therefore, choosing a cut-off threshold between abnormal and normal can be regarded as arbitrary (Bostrom & Roache, 2007; Freitas, 2007; Harris, 2007). Bostrom and Roache have argued in connection with this that 'to define abnormality as falling (say) one or two standard deviations below the population average seems to lack any fundamental medical or normative significance' (2007, p. 2).

Considering all this points, we can say that the distinguishing line between health and ill health is no more than an academic idealisation, rather than a fact of life. To quote H. Tristram Engelhardt, 'what counts as health and disease for humans depends upon very complex judgements concerning suffering, the goals proper to humans, and for that matter the form of appearance proper to humans' (Engelhardt, 1976, p. 102). Thus, using health and disease as the concepts to distinguish therapy from enhancement is not that useful; in particular, because these concepts are already immersed in the ongoing debate concerning the goals and limits of humans, and the role of medical practice in achieving them.

The changing goals of medicine

Medicine has always been interventionist in nature, and its beginnings can probably be traced back to those instances when individuals suffering from injury or ill health asked other individuals (for instance an elder member of the community or shaman) to help them with their pain and afflictions.[22] The goals of medicine in a traditional sense were to cure diseases when possible; to reduce suffering; and to alleviate pain and other manifestations of diseases (Hook, 2007; Illich, 1975). Thus, traditionally, medicine's main goal has been mostly focused on care rather than on curing: 'Cure when possible, care always' (Hook, 2007, p. 345). However, these goals have changed, and now the goals of biomedicine are focused on treating disease and preventing risks to health (Engelhardt, 1976). In addition, the fact that disease and health, as mentioned above, are already controversial and changing concepts makes them problematic as the basis of medicine's goals and in discussing the moral relevance of therapy versus enhancement interventions.

As our understanding of health, disease and the goals of medicine have changed so have the professional specific moral duties we ascribe to physicians (Hook, 2007; Ravelingien, Braeckman, Crevits, De Ridder, & Mortier, 2009; Wolbring, 2005) and the role of patients. For the latter their role has changed from individuals suffering from ill health to consumers of health. Changes in our understanding of health, disease and the goals of medicine have also promoted an increasing use of 'lifestyle' drugs among healthy individuals (Racine, 2010); and exacerbated the medicalisation process. All of these changes could indeed represent a challenge to the integrity of medicine as a profession, and contribute to puzzlement around normality, treatment, enhancement and medical need. To quote Gerald McKenny:

> In the absence of such a framework [to evaluate and criticize the commitments of modern medicine], the commitment to eliminate all suffering combined with an imperative to realize one's uniqueness leads to cultural expectations that medicine should eliminate whatever anyone might consider to be a burden of finitude or to provide whatever anyone might require for one's natural fulfilment.
>
> (McKenny, 1997, p. 20)

I quote this at some length because advances in emergent technologies with a biomedical purpose in mind have taken away the familiarity and uncontroversial nature that we used to grant to medicine. The use of emergent technologies within biomedicine has come along with promises about treating diseases better and in some cases even ending them; offering more effective ways to prevent risks to health; and the possibility of attaining health with fewer side effects, in a less burdensome and cheaper way. Since these technologies are merely seen as an instrument to attain the (supposedly) widely shared goals of medicine—which are seen mostly as familiar and uncontroversial—their use is regarded as justified and not morally controversial (Boenink, 2009).

However, it is worth questioning whether biomedical goals are really so familiar and uncontroversial. Artificial organs, xenotransplantation and preimplantation genetic diagnosis are just some examples of how new technologies have changed our views about our bodies and reproduction. Technology shapes to a great extent the ways in which we experience the world and ourselves, and the way we act in regards to both (Boenink, 2009; McKenny, 1997; Postman, 2006). Considering the impact that emergent technologies could have for biomedicine, better ways are needed in which to deliberate about the challenges that these advances are likely to bring forward. Therefore, before we can even attempt to use the concepts of disease, health and the goals of medicine to draw a moral distinction between therapy and enhancement, an urgent revision is needed of the way these concepts are conceived in our societies.

The biomedical paradigm: Therapy versus enhancement

The concepts discussed in the previous section are grounding concepts for what is understood as therapy. Therapy, within biomedicine, is generally associated with maintaining and restoring health or becoming well, thus in accordance with the traditional goals of medicine. However, from our previous discussion it is plausible to say that biomedicine also provides a model case of how science and technology can be ordered beyond the 'traditional' goals of medicine to eliminate whatever anyone might consider a burden of finitude or to achieve whatever anyone might consider as bodily excellence (McKenny, 1997);[23] that is to say, towards instances where there is no clear therapeutical benefit. These instances are often regarded in the literature as instances of a state better than well or beyond therapy (Coenen et al., 2009; Khushf, 2004; PCB, 2003; Sandel, 2004). This conceptual distinction is known as the therapy-enhancement distinction and it is the basis for the biomedical paradigm, the predominant view on human enhancement in the contemporary Western tradition. The following section reviews and discusses these two concepts—therapy and enhancement—as used within the biomedical paradigm, as well as mentioning the key features of the biomedical paradigm.

Biomedical enhancement

Under the biomedical paradigm human enhancement is generally understood as any 'medical' intervention that goes beyond the scope and goals of therapy, and in some cases even biomedicine (Juengst, 1998; PCB, 2003; Wolpe, 2002). There are different basic concepts used to describe therapy. A commonly used concept is that of normality. Norman Daniels has used the 'quasi-statistical concept of normality' (Daniels, 1992, 2000, 2008) to argue that interventions 'designed to restore or preserve a species-typical level of functioning for an individual should count as treatment', whereas those that 'would give individuals capabilities beyond the range of normal human variation' are enhancement (Daniels, 1992, p. 46). Others, such as Eric Juengst, have suggested a mixture between health and normality concepts, arguing that therapy involves interventions that deal with pathologies that compromise health or reduce one's level of functioning below the species-typical or statistically normal level, whereas enhancement interventions aim at improvements that are not health-related (Juengst, 1997). Another health-based therapy definition is the one stated by the Bush-era President's Council on Bioethics (PCB) in its report 'Beyond Therapy', in which therapy is defined as interventions 'to treat individuals with known diseases, disabilities, or impairments, in an attempt to restore them to a normal state of health and fitness' (PCB, 2003, p. 13). Both the normality and health-based definitions endure, as the recent President's Commission defines enhancement as 'pharmacological and technological interventions meant to improve mental and physical capacities beyond normal functioning' or used for a 'non-diseased capacity' (PCSBI, 2014). Most accounts of therapy follow one of these definitions or share their main idea. The term 'therapy' is also used to describe interventions involving ideas connected with curing, reversing, restoring, halting or ameliorating disease, disability, ill health and making the patient *well*.

Consequently, we end up with a broad range of possibilities for what counts as an enhancement intervention. For instance, human enhancement has been described as any intervention that (a) is aimed at making ourselves 'better than well' (Sandel, 2004); (b) boosts our capabilities beyond the species-typical level or statistically normal range of functioning for an individual (Daniels, 1992, 2000; Juengst, 1997); (c) increases our general capabilities for human flourishing (Chan & Harris, 2008; Degrazia, 2005; Savulescu, 2006); or (d) improves human form or functioning beyond what is necessary to sustain or restore a healthy state (Bostrom & Sandberg, 2009a; Juengst, 1998). Taking into consideration the main ideas put forward by these understandings, biomedical human enhancement is understood here as:

> **Biomedical human enhancement**: any intervention aimed primarily at the improvement of one or more core capacities of an individual beyond what is therapeutical necessary to sustain or restore health as conceived within biomedicine.

The paradigm

The biomedical paradigm assumes that it is possible to draw a line between therapy and enhancement, and that such a line should be drawn because the distinction is needed to inform ethical debate (Fukuyama, 2002; Habermas, 2003; PCB, 2003; Sandel, 2004). On the one hand therapy is seen as responding to genuine medical needs and as such generally regarded as uncontroversial; on the other hand enhancement interventions are regarded as going beyond being well and are thus regarded as ethically controversial. These differences are the framework of the well-known therapy-enhancement distinction.

In what follows it will be considered that enhancement interventions, under the biomedical paradigm, are always taken to be morally controversial. The presumption held by this paradigm for considering enhancement as something biomedicine should not get entangled with, is that enhancement goes beyond the goals of medicine, beyond making an individual healthy. However, as we have already discussed, trying to define enhancement using the concepts of biomedicine is not without ambivalence.

The particular stand of this paradigm on human enhancement, in which enhancement is regarded as intrinsically morally wrong, locates it on the bioconservative side of the biopolitical spectrum. The different strands of contemporary bioconservative views can be traced to multiple origins, such as the Greek concept of hubris (deep personal intuitions), religious views, Luddite workers' revolt against industrialisation, various Continental philosophers' critiques of technology and the rationalistic mindset that accompanies modern technoscience, and objectors to the consumerist rat race. Thus, the variety of arguments against enhancement that one finds under different strands of bioconservative views is not surprising. A general bioconservative view towards human enhancement technologies, such as nanotechnology, is the need for a national or international ban. Accordingly, the biomedical paradigm has a strong emphasis on the role of regulation for restricting enhancement research and applications. However, recently more diverse views about the permissibility of biomedical-based human enhancement have started to emerge. Take for instance, the article published in *Nature* by a group of bioethicists and scientists (Greely et al., 2008) supporting the responsible use of cognitive-enhancing drugs by the healthy. Another more recent example is the report by the American Academy of Neurology, crafted by the Ethics, Law and Humanities committee, which concluded that prescribing or recommending products to healthy adults is generally acceptable (Larriviere, Williams, Rizzo, & Bonnie, 2009). However, these are not the views characterising the biomedical paradigm as constructed here.

One more important feature of this paradigm is that the ultimate goal of biomedical enhancement is to end disease, and become as healthy as one can be within species-typical limits. Now, even though there might be interventions, such as immunisation programmes, in which it is possible to argue that human enhancement is aimed at improvement of the species (Coenen et al., 2009), most interventions within this paradigm are focused on individuals. Accordingly, the values it promotes are individualistic and based on the concept of the liberal individual, in which individuals are abstract, rational, self-sufficient and isolated beings (Held, 2001, 2006). Furthermore, this liberal individualistic framework is focused on self-interest, individual well-being and freedom of choice, as well as respect for the autonomy of individual persons.

Finally, under this paradigm enhancement interventions are seen as *improvement of*. In the HLGE report it was suggested to distinguish between enhancement *for* and enhancement *of* (Nordmann, 2004), a distinction also known as the software-hardware approach. The hardware view refers to enhancement *of* and is committed to hardwired technology, where hardwiring can be implemented at different levels of the mind and body, such as molecular design or neuroimplants. The software view refers to enhancement *for* and does not seek to interfere directly with the brain or body's architecture, but rather to exploit causal pathways that change how information is accessed and processed (Nordmann, 2004). Under this paradigm, the use of emergent technologies for enhancement interventions follow a hardwired approach, being aimed at changing and interfering directly with the brain and the body.

The role of emergent technologies in the biomedical paradigm

New scientific discoveries have promoted a way of understanding life at the molecular level. This 'style of thought', as Nikolas Rose puts it, has put great emphasis on the role that the nanometre-scale world—together with the various possibilities it enables (such as better identification, isolation, manipulation and mobilisation)—could have for human health (2007). This style of thought underlies the use of emergent technologies within this paradigm. Evidence of this is the view that emergent technologies will allows us to gain molecular knowledge of the human body and enable the molecular tools for their manipulation, or that the human body can be reduced to algorithms and thus can be reprogrammed. Another view is that the use of these technologies in medicine will achieve faster and more cost-effective interventions, with higher performance in terms of sensitivity, resolution, reliability, robustness, integration and reproducibility. These potential benefits of emergent technologies for medicine have been widely recognised in both the scientific literature and popular media (Boenink, 2009; Bouamrani et al., 2005; Combs, 2006; Freitas, 2005a, 2007; Hodgins et al., 2008; Keravnou, Garbay, Baud, & Wyatt, 1997; Malsch & Nielsen, 2010; National Research Council, 2008; Roco & Bainbridge, 2002; Ruder, Lu, & Collins, 2011; Tomellini, Faure, & Panzer, 2006).

In some cases emergent technologies in medicine are portrayed as a magic bullet to improve the effectiveness of therapies and diagnosis of disease, for instance, by creating interventions that are less dependent on refrigeration, less dependent on highly specialised research infrastructures, or are less invasive. In the case of disease diagnosis the promise is to enable earlier diagnostics with greater reliability, and to improve the prognosis to reduce over/under-treatment. These technologies are also envisioned as a way to combat disease in a more efficient, cheaper and safer way, as well as enabling 24/7 monitoring of bioparameters and personalised care. In other cases these advances are questioned and are thought as raising problems connected to fair distribution and safety. Regardless of which approach we take, it cannot be refuted that emergent technologies will influence the content and formation of biomedical knowledge, tools, practice and values, and consequently the biomedical paradigm of human enhancement.

Another feature of the use of these technologies under the biomedical paradigm is that it is focused mostly on individualistic interventions. While certain applications of these technologies have started to focus on public health outcomes, such as fighting infectious diseases, chronic diseases and vaccine programmes; the values motivating such interventions are still predominantly individualistic.

There is an expanding list of biomedical enhancement interventions (Bostrom & Sandberg, 2009a; Farah & Wolpe, 2004), including smart drugs, brain implants, genetic manipulation and brain stimulation. Some examples of how emergent technologies are being used within the biomedical paradigm are the following: to create new medical material and systems; to achieve higher integration density of the electronic circuitry in smaller systems;[24] to achieve better solubility, better distribution within the body, improved selectivity and specificity of targets (be they molecules, cells, tissues or organisms); to achieve more favourable pharmacokinetics (such as reducing toxicity, damage/side-effects/adverse effects and reabsorption); to develop more reliable biosensors and measurement systems;[25] to develop systems for delivery of certain molecules or nutrients for cells and removal of waste products; and to help early diagnosis, detection and monitoring.

Particular applications of these technologies include improved brain machine interfaces (Berger et al., 2005; Donoghue, 2002, 2008; Nicolelis, 2003) and sensory substitution systems (Bach-y-Rita & Kercel, 2003; Lenay, Gapenne, Hanneton, & Marque, 2003). Connected to the latter, there are projects focused on how to enable people to see by using their tongue (Bach-y-Rita & Kaczmarek, 1998).

Pharmacological interventions

It is already possible to observe a wide range of drugs that are used for enhancement purposes. The most common ones include hormones, stimulants, nutrients and neuromodulators (Farah, 2010; Farah & Wolpe, 2004; Greely et al., 2008; Turner & Sahakian, 2006). Modafinil and Ritalin have been paradigmatic cases in the discussion of pharmaceuticals and cognitive enhancement (Cahill & Balice-Gordon, 2005; Ilieva & Farah, 2012; Maslen, 2014; Repantis, Schlattmann, Laisney, & Heuser, 2010; Schermer, Bolt, de Jongh, & Olivier, 2009), Prozac for affective enhancement (Degrazia, 2000; Kramer, 1993), oxytocin for prosocial improvement (Kosfeld, Heinrichs, Zak, Fischbacher, & Fehr, 2005; Weisman, Zagoory-Sharon, & Fregni, 2012).

Under the biomedical paradigm, pharmaceutical interventions are not necessarily for impaired or at-risk 'patients', but rather for lifestyle uses (also referred to as 'off label uses' or the misuse of prescription drugs) (Farah, 2010; National Research Council, 2008; Racine, 2010; Racine & Forlini, 2010). Interestingly, most documented uses of off-label enhancement drugs occur within the academic setting, either by students or academics themselves (Farah, 2010). Hence, even though the long-term effects of enhancement interventions like this in healthy individuals remain largely unknown (Chatterjee, 2004), their potential significance in the field of cognitive enhancement has been acknowledged by researchers (Greely et al., 2008).

Genetic interventions

Advances in genetics have made it possible to relate specific genes to human performance and mental function (Green et al., 2008). There are techniques that make it possible to turn genes on and off and to delete specific genes in specific brain regions, but most of these techniques have only been tested in animal models. A recent example is the research of professor Li-Huei Tsai and colleagues, which shows that the gene *HDAC2* regulates the expression of many genes implicated in brain plasticity and memory formation, suggesting that it is one of the major targets involved in eliciting memory enhancement (Guan et al., 2009). However, there are many other unknown genes that are likely to contribute to different human traits and functions; but the idea of manipulating specific genes to achieve desired physical or behavioural traits is not in the near future. A more likely avenue for genetic enhancement is that of gene replacement, in which pre-engineered genes could replace genes that are known to contribute to unfavourable attributes and substitute these with more favourable and desirable traits (Sade & Khushf, 1998).

Neural interfaces and implants

More complex interventions include neural interfaces, such as brain-computer interfaces (BCI) and brain-controlled interfaces. While these interventions were originally 'designed to restore control, communication, and independence to persons with paralysis when the motor control structures are disconnected from muscle output' (Donoghue, 2008, p. 511), they are now also used or envisioned for purposes beyond therapy. The aim is that in the near future they will be fully integrated with the brain and capable of interfacing with several other devices (National Research Council, 2008; Task group summaries, 2007).

One example of emergent technologies in the area of neural interfaces is the use of carbon nanotubes attached to specific neurons for enhancing their natural signal-processing capabilities (Johnson, 2008). Researchers also hope to use carbon nanotube-based circuitry to improve neural implants and to create the next generation of thought-controlled prosthetics (Donoghue, 2002, 2008).

Brain stimulation

Brain stimulation techniques also have the potential for human enhancement applications (Bioethics, 2013; Desmond & Pascual-Leone, 2006; Gagnon, Schneider, Grondin, & Blanchet, 2011; Gladwin, Uyl, Fregni, & Wiers, 2012; Hamilton, Messing, & Chatterjee, 2011; Javadi & Cheng, 2013; Luber & Lisanby, 2014; Suthana et al., 2012). Transcranial Magnetic Stimulation (TMS) and transcranial Direct Current Stimulation (tDCS) are two brain stimulation technologies in which the individual undergoing the procedure remains conscious and as such are regarded as a minimally invasive form of brain stimulation. On the other hand, Deep Brain Stimulation (DBS) is a more invasive technique, as it requires brain surgery.

The idea behind TMS and tDCS is that the current induced in the brain triggers molecular changes, modulating cortical excitability (tDCS) and spontaneous firing activities in the stimulated region (TMS). In the case of TMS trains of magnetic pulses are administered by electromagnets, creating localised electrical currents in the brain that activate certain neurons or stimulate the production of certain neurotransmitters, which change cerebral activity when applied in a repetitive manner. TMS studies have shown improved performance in language-related abilities, such as naming facilitation (Mottaghy et al., 1999; Mottaghy, Sparing, & Töpper, 2006), various complex motor learning tasks (Kim, Park, Ko, Jang, & Lee, 2004; Kobayashi, Hutchinson, Theoret, Schlaug, & Pascual-Leone, 2004), visuospatial processing (Hilgetag, Théoret, & Pascual-Leone, 2001; Walsh, Ellison, Battelli, & Cowey, 1998), perceptual abilities (Gallate, Chi, Ellwood, & Snyder, 2009; Snyder, 2009; Snyder et al., 2003) and modulating social cognition (Knoch, Pascual-Leone, Meyer, Treyer, & Fehr, 2006; Lo, Fook-Chong, & Tan, 2003; Luber, Fisher, Appelbaum, Ploesser, & Lisanby, 2009; Young et al., 2010).

In the case of tDCS a weak direct current, less than one-tenth of the current flowing through the earbuds of an iPod, is applied to the scalp via two saline-soaked sponge electrodes. The current applied polarises the underlying brain tissue with electrical fields, and different studies have suggested its potential for enhancing working (Fregni et al., 2005; Ohn et al., 2008) and declarative memory (Marshall, 2004) as well as certain forms of learning, such as naming facilitation and visual-motor learning (Antal, Nitsche, & Kincses, 2004; Bullard et al., 2011; Fertonani, Rosini, Cotelli, Rossini, & Miniussi, 2010; Floel, Rösser, Michka, Knecht, & Breitenstein, 2008). Evidence for the enhancement of more general complex problem-solving abilities via tDCS is limited, but intriguing. This includes areas such as complex verbal associative thought (Cerruti & Schlaug, 2008), planning ability (Dockery, Hueckel-Weng, Birbaumer, & Plewnia, 2009), numerical competence (Cohen Kadosh, Soskic, Iuculano, Kanai, & Walsh, 2010) and problem-solving (Chi & Snyder, 2011; Snyder, Ellwood, & Chi, 2012).

In DBS electrodes are implanted in the brain, together with an electrical pacemaker, implanted under the patient's clavicle, that controls the settings of the brain implant. While DBS has the advantage to enable the stimulation of more specific areas of the brain compared to its minimally invasive counterpart it is also more invasive. The benefits-risk ratio of DBS is the reason why to date only patients with undoubtedly burdensome disorders are treated by it; thus there are currently no efforts to use DBS in healthy individuals for enhancement. The few cases of enhancement via DBS have been the result of an unexpected outcome of the medical application. For instance, Hamani and colleagues (2008) noticed that their patient who was been treated for morbid obesity reported increased recollection (but not familiarity-based recognition). Since then other studies have suggested that DBS could improve certain memory functions (Hescham et al., 2012; Hu, Eskandar, & Williams, 2009; Suthana et al., 2012). DBS has also been reported to enhance mood (Schermer, 2011; Synofzik, Schlaepfer, & Fins, 2012) and produce behavioural changes. Thus, considering these results it is not far-fetched to think that in the future some people might try to use DBS for stimulating their personalities and enhancing their mood.

While these examples illustrate the possibilities that brain stimulation has for human enhancement, more evidence is needed before any clear conclusions can be drawn (Bioethics, 2013). The last section of this chapter will analyse the main arguments used within the biomedical paradigm to reject enhancement, some of the main problems the paradigm faces and finally the value this paradigm has.

The ethics of biomedical human enhancement

This section analyses some of the most recurrent arguments used against biomedical human enhancement interventions. As mentioned earlier, given that this paradigm considers it to be morally wrong to pursue enhancement, most of the arguments will be in line with those used by the bioconservatives. Other arguments that are not unique to this paradigm, such as identity, will be discussed in Chapter 5.

The arguments

Naturalness

One of the most common arguments used by people against human enhancement is connected to the idea that enhancement interventions are somewhat 'unnatural' or that they trespass on the natural. This view assumes that the natural is good. Furthermore it considers the acceptance of our *natural* limitations as the only way to live a proper human life. In connection to this, Michael Sandel has said that 'medical intervention to cure or prevent illness or restore the injured to health does not desecrate nature but honours it' (2004). On similar lines, Jürgen Habermas has stated: 'what is so unsettling [about enhancement interventions] is the fact that the dividing line between the nature we are and the organic equipment we give ourselves is being blurred' (2003, p. 22). Interfering with nature is seen as blurring the line between chance and choice, which has been the backbone of our value system and morality (Buchanan, Brock, Daniels, & Wikler, 2001; Dworkin, 2002).

Others have focused on arguing that if we were to lost sight of the difference between human-engineered and not-human-engineered things, the very ends we desire might become separated from ideas about what is humanly superior, and therefore humanly worth seeking or admiring. This view assumes that using technology to radically enhance the human condition is inherently wrong or morally suspect on the grounds that it will cause us to lose our humanity (Dworkin, 2002; Fukuyama, 2002; Kass, 1997; Peters, 2007). In some instances this intuition is based on religious views, such as 'playing God'. The argument here roughly states that since a supreme force created life, we would be playing or challenging God, or messing with his creation, by trying to change the nature of the human condition (Dworkin, 2002; Peters, 2007; Sandel, 2004). At other times the intuition is based on a sort of respect towards evolutionary design or a more visceral reaction, such as repugnance (Kass, 1997). Connected to this, Leon Kass has argued that the wisdom of repugnance might be 'the only voice left that speaks up to defend the central core of our humanity' (Kass, 1997, p. 20).

Autonomy and freedom

Another common argument against human enhancement is that it is a threat to our autonomy and freedom. Both notions are complex and controversial, but the basic idea underlying them is that one should have the various aspects of one's life under one's control and should be able to decide for oneself without being shaped by external influences. Accordingly, arguments against human enhancement on autonomy and freedom grounds assume certain alienation or instrumentalisation of the human condition, in which somehow we will no longer be in charge of our own lives, thus not being able to decide for ourselves, to take responsibility, make our own choices and take control over who we will become. The culmination of the fear underlying these arguments becomes visible in the idea of a diminished society of spectators just as the one described in Aldous Huxley's 1932 novel, *Brave New World*. In connection with this, Kass has argued that '[u]nlike the man reduced by disease or slavery, the people dehumanized à la Brave New World are not miserable, don't know that they are dehumanized, and, what is worse, would not care if they knew' (1985, p. 35). Habermas has argued that such a stage of inhuman level of predestination affects our 'autonomous conduct of life and moral understanding' (2003, p. 52), excluding us from the 'spontaneous self-perception' (p. 63) of being the authors of our lives. Habermas sees this ethical self-understanding of the species as crucial for our capacity to recognise one another as autonomous persons.

Most arguments about autonomy and freedom also assume that the capacity of being oneself requires the individual to be in his or her own somatic being (referring to our bodies in a broad sense). While related to the first argument connected to naturalness, this assumption stands on its own as there is a distinction, even if not a clear one, between 'being a body' and 'having a body' (Plessner, 1961). To be oneself requires us to understand (or at a minimum to be engaged in trying to understand) what our somatic being is, in spite of what we add to it.

Human dignity

Another concern comes from the idea that enhancement interventions are threats to human dignity. In this regard most people within this paradigm take for granted the importance of human dignity, even though the term is highly abstract and contested. For instance, Francis Fukuyama has argued that:

> Denial of the concept of human dignity—that is, of the idea that there is something unique about the human race that entitles every member of the species to a higher moral status than the rest of the natural world—leads us down a very perilous path.
>
> (2002, p. 160)

Depending on the specific view of human dignity that one holds, this argument has many versions. For instance, according to some religious views human dignity is a gift from god or the gods. Secular understandings of dignity include a philosophical/legal conception and a biological conception. The former is based on a Kantian notion of 'personal dignity' in which dignity indicates a set of capacities and qualities that we ascribe only to persons, such as rationality, self-awareness, ability to communicate and self-determination, to mention a few; whereas the latter refers to the biological nature of human beings, regardless of their developmental status. Both conceptions are used, sometimes in a very entrenched way. Fukuyama sees human dignity as the consequence of an 'essential quality that has always underpinned our sense of who we are and where we are going, despite all the evident changes that have taken place on the human condition through the course of history' (2002, p. 101). He describes this essential quality as something that is worth a certain level of respect and that 'cannot be reduced to the possession of moral choice, or reason, or language, or sociability, or sentience, or emotions, or consciousness, or any other quality that has been put forward as a ground for human dignity' (2002, p. 171); rather it is all of these qualities coming together that makes this an essential human quality.

According to this view, introducing technologically enhanced individuals into society threatens human dignity, because this results in the loss of moral status that 'ordinary' humans currently possess. This view presupposes that enhanced individuals do not have equal dignity to unenhanced humans.

Social disruption

Another argument used under this paradigm against human enhancement, and one closely related to the previous one, is the negative social consequences that it can bring, such as (a) a potential irreversible and radical change in the human species (Fukuyama, 2002; Habermas, 2003) and (b) a new form of eugenics (Sandel, 2004).

(a) This rests on two main assumptions. First, it assumes that differences between enhanced and unenhanced human beings would create problems of discrimination and possible social conflict, in which enhanced humans could harm unenhanced ones or even plan to eliminate them (Fukuyama, 2002). Historians have argued that whenever there has been a struggle of races, the surviving race is the one which has proved to be physically and mentally stronger.

If this view is correct, then the survival of unenhanced humans can indeed be questioned. The second assumption held by those who are afraid that enhancement could create significant differences between enhanced and unenhanced individuals, is that our notions of human rights and political equality are based on a resemblance between human beings as morally equal beings with the same dignity. Thus, if this common framework is dissolved, social and political instability is likely to take place.

(b) Connected to the issue of promoting a new kind of eugenics, arguments have focused on the idea that enhancement will reinforce stereotypes, stigmatisation and discrimination (Wolbring, 2006, 2008a), and that people would be coerced to undertake enhancement interventions. This takes us again to issues connected with the idea of promoting a *Brave New World* scenario.

Authenticity

Authenticity is another common argument against human enhancement. There are two interrelated ideas in this line of argumentation: cheating and the good life. With regard to cheating there are two parts to the argument. On the one hand, by using enhancement interventions people have an unfair advantage over others, thus not being authentic to the ethos of the activity itself. On the other hand, cheating can also be about cheating oneself out of the full value of an activity (Goodman, 2010). For instance, Maartje Schermer (2008) uses the example of someone who claims to be a 'mountain climber' but takes a helicopter to the summit. This individual is not really harming a competitor or breaking an explicit rule, but he or she is cheating himself or herself out of the rewards and challenges of an unaided ascent, such as personal growth or the development of wisdom, both of which we generally associate with a good life. This line of argument implies that accomplishment is deserved and worthwhile only when there is sacrifice, hard work, effort and in some cases even pain and suffering involved.

Schermer is not the only one who shows concern with enhancement on the grounds that it makes activities less difficult, and that it makes them less meaningful. The PCB's report argues on similar lines, stating that '[w]e are deforming also the character of human desire and aspiration, settling for externally gauged achievements that are less and less the fruits of our own individual striving and cultivated finite gifts' (PCB, 2003, p. 150). Sandel (2004) is also concerned with the dispositions that enhancement promotes, not just the dispositions it expresses. This particular line of argument about the message that a certain practice conveys is also known as the expressivist objection (Brock, 2009).

From this perspective it can be said that enhancement turns our acts from something we achieve into something that happens to us, thus corrupting our character and making us inauthentic to our own selves. This authenticity argument connects to the idea of the good life, inasmuch as we believe that the good life requires oneself to be true or authentic to oneself (Taylor, 1991) and to one's own accomplishments. Thus, the good life is only accessible when one does not cheat oneself.

The following section will explore why the therapy-enhancement distinction, regardless of the plausibility of some of the moral arguments it suggests, might not be the most suitable paradigm on which to base our human enhancement practices.

The issues

The following list, which by no means is comprehensive, summarises the main reasons for considering the therapy-enhancement distinction as problematic for helping us to deliberate about the moral implications of human enhancement.

Its foundation concepts

As discussed at the beginning of this chapter, the basic concepts on which the therapy-enhancement distinction is grounded—such as health, medical necessity, non-disease states, normal or natural—are themselves ambivalent, dynamic and socially constructed concepts. For example, many contemporary biomedical practices already go beyond the traditional goals of biomedicine and are no longer regarded as morally reprehensible, including contraceptive devices, fertility treatments and some cosmetic procedures. This leaves us with no clear reference point that discriminates between therapy and enhancement in practice.

How to draw moral boundaries in these shifting areas is definitely not as straightforward as some scholars have implied. Surely, while some therapies should be regarded as morally dubious, not all enhancement interventions need to be regarded as such. Furthermore, early diagnosis and presymptomatic therapies will erode even more the distinction that the biomedical paradigm tries to sustain between therapy and enhancement interventions, because by definition a presymptomatic intervention entails that no symptom has yet taken place (Miah, 2008).

Definition

Another main objection against the therapy-enhancement distinction is related to the vagueness of the concept *enhancement* itself. It has been argued that it is not possible or realistic to draw a distinguishing line between therapy and enhancement or that it is blurry at best (Bostrom & Savulescu, 2009; Harris, 2007; Juengst, 1997; Levy, 2007; Parens, 1998). The assumption made by the biomedical paradigm regarding the idea that enhancement can be differentiated—either qualitatively or quantitatively—from therapy, can be challenged by arguing that therapy and enhancement do not need to be incompatible or mutually exclusive (Harris, 2007, 2009). Thomas Murray has argued, for instance, that therapy and enhancement overlap because both are built upon the 'intrinsic healing processes of the body and mind with the aim to restore the body to a natural, balanced state' (quoted in *Society for Neuroscience*, 2005, n.p.). The paradigmatic example around this issue has been vaccines, which generally work by enhancing the immune system, but are generally regarded as within the scope of therapeutical interventions (Daniels, 1985, 2008). Vaccines are interventions that do not restore or heal any previous health condition; rather they are interventions focused on reducing the probability of developing certain diseases. If we consider their mechanism of action, we can even say that they are more than just preventive interventions. For instance, John Harris has argued that 'measures that protect humans from things to which they are normally vulnerable, or which prevent harm to that individual by operating on the organism by affecting the way the organism functions, are necessarily also enhancements' (2009, pp. 152–153). In addition to vaccines there are other instances of interventions that are in the border of therapy and enhancement, such as organ transplantation, some reproductive technologies and brain stimulation techniques. Furthermore, there are areas of human life that in the past were not considered in need of treatment, such as erectile dysfunction after a certain age, hair growth, birth control, better memory or removal of wrinkles.

Another reason that makes a clear cut distinction between therapy and enhancement problematic is that the same intervention can be therapeutic for a certain individual while being an enhancement for another (Harris, 2007). Given the fact that human capacities fluctuate, not only within a population but also within the lifespan of an individual, also makes it unclear whether or not an intervention that restores capacities individuals used to have in the past should count as therapy or as enhancement. Finally, it can be argued that the distinction to be captured by using the terms 'therapy' and 'enhancement' is not one between two completely different things, but rather one that points out the extremes of a continuum (Selgelid, 2007).[26] Patrick Lin and Fritz Allhoff (2008a) have used the 'paradox of the heap' to argue that even though there might not be a 'clear distinction between a heap of sand and a less-than-a-heap or even no sand at all' (p. 2), it would be the wrong conclusion to say that there is no difference between them. Surely when thinking of therapy and enhancement as the end points of a continuum the distinction might be obvious and clear, but as we move through the different possible cases along the continuum the distinction stops being so obvious and clear. It is important to keep this in mind because the more challenging ethical issues regarding human enhancement are to be found along the continuum rather than at the end points.

Dualism

The biomedical paradigm still relies on a form of dualism, either the mind-body distinction or the patient-expert one (Bennett & Hacker, 2003; Kirmayer, 1988; Scheper-Hughes & Lock, 1986). This has led to a view of the body and the brain as biochemical machines, in which our bodies and its features can be explained in mechanistic terms, rationalised for efficiency, and intervened and fixed through technological interventions (such as implants or pharmaceuticals); a view that distances itself from the patient's emotions and subjective experience (Foucault, 1990; Kirmayer, 1988; Scheper-Hughes & Lock, 1986). For instance, Foucault and Kirmayer have argued that the architecture of hospitals, the gowns and even medical language are ways used by biomedical practice to distance the patient from their bodies and their subjective experiences.

Needs and preferences

The therapy-enhancement distinction argues that medical needs should be prioritised over individual preferences. According to Sabin and Daniels (1994), medically necessary interventions are those that effectively treat disease and disability or ameliorate conditions deriving from them. However, what counts as a medical need and the ways to address it is not a straightforward matter, particularly as new forms of defining disease, such as the one suggested by Freitas, emerge and social processes such as medicalisation expand. Moreover, in the kind of globalised and market-driven world in which we live, we face the problem that too many things become needs, but also too few things are considered needs (Daniels, 2008). Even if we could agree on what counts as a medical need, in a world of limited resources we still need to prioritise between competing needs (Daniels, 2008). There are two main problems related to this. On the one hand there is the fair chances/best outcome problem, which has to do with the idea of whether resources allocation should consider the likelihood that someone will benefit from those resources or not. And on the other hand, there is the aggregation problem, which confronts us with the question of whether 'minor benefits to larger numbers of people... outweigh significant benefits to fewer people' (Daniels, 2008, p. 114).

The issues presented here are the main inherent issues in the biomedical paradigm. In the next section the valuable aspects of the paradigm will be discussed.

The value of the biomedical paradigm

Even with a substantial number of flaws, the biomedical paradigm has been the predominant human enhancement paradigm. There are a couple of reasons for thinking that even if this paradigm is not the best it could still offers valuable insights. On the one hand, most of the pro-enhancement advocates have used the therapy-enhancement distinction to argue for enhancement (Allhoff, Lin, Moor, & Weckert, 2010). So it seems to be a practical (or convenient), rather than useful, distinction for both sides of the debate. That explains why some people have argued that the distinction should be maintained for pragmatic reasons, in particular in the health policy context (Coenen et al., 2009; Selgelid, 2007). Certainly, we can still question how practical a distinction that is so ambivalent and biased can be, or practical for whom.

On the other hand, given that most people from liberal Western societies accept an understanding of health in accordance with biomedicine, the distinction made under the biomedical paradigm fits general intuitions. Even if there is no clear line, or no line at all, that differentiates therapy from enhancement, the distinction has helped to avoid the view that everything is acceptable. Nonetheless, it might be about time to start thinking in new ways to assess the permissibility of human enhancement when using emergent technologies. Thus, to the extent that we are still working out better ways in which to reflect upon human enhancement, the therapy-enhancement distinction can still play an important role in arguing about the permissibility of certain interventions (Allhoff et al., 2010).

Conclusion

This chapter introduced the biomedical paradigm of human enhancement based on the therapy-enhancement distinction. The first section of this chapter discussed the main concepts and ideas underlying this paradigm—disease, health and the goals of medicine. The main features of this paradigm as well as the role that emergent technologies play in human enhancement within this paradigm were highlighted. The second section of this chapter presented the biomedical paradigm's ethical stand towards enhancement by showing five different arguments against it—naturalness, autonomy and freedom, human dignity, social disruption and authenticity. The issues and challenges this paradigm faces as well as the value it has within the human enhancement debate were also discussed.

3
The Transhumanist Paradigm

Transhumanism and the posthuman

The previous chapter dealt with what today still remains the dominant human enhancement paradigm; however, other views on human enhancement have started to emerge along with social, technological and political changes across the world. One recent but rapidly growing social and cultural movement that is shaping a different view on human enhancement is transhumanism. Thus, the paradigm to be explored in this chapter is referred to as the transhumanist paradigm. In order to better grasp the values and perspectives held by this paradigm, this chapter starts with an overview of the different concepts that have shaped it—transhumanism, posthumanism and the posthuman.

Transhumanism

The human enhancement debate is one where issues regarding the features and meaning of being human as well as issues about transcending human limits are constantly discussed. New technological advances blur more and more the parameters we use to distinguish ourselves from our machines. In addition our assumptions about the uniqueness of humans are put into question, with new discoveries framing key human features to mechanical, predictable and controllable processes.

The quest for transcendence and improvement of our human condition is probably as old as humankind itself. Different epochs, different cultures and even different individuals might have different views about the scope and path of the quest, but it can be argued that it is a quest that has defined humankind. In the Western world, this quest took a new perspective during the Age of Enlightenment when Francis Bacon proposed to use science as a means to achieve mastery over nature, including our own. Likewise, during the Renaissance, with its rational humanism, man saw the possibility of using technology to improve the human organism. These have been regarded among the first sources of inspiration for transhumanism. Other important sources of inspiration in shaping transhumanist ideas have been the work of the German philosopher Friedrich Nietzsche,[27] English writer H. G. Wells and English historian and philosopher Winwood Reade (Sorgner, 2009; Coenen, 2014).

All these writers share with transhumanists' writings a dynamic view of human values and nature as well as ideas about overcoming or transcending the human condition. To quote Nietzsche:

> Man is a rope stretched between the animal and the Superman—a rope over an abyss. A dangerous crossing, a dangerous wayfaring, a dangerous looking-back, a dangerous trembling and halting. What is great in *man is* that he is a *bridge and not a goal*: what is lovable in man is that he is an *over-going* and a *down-going*.
>
> (Nietzsche, 1999 [emphasis added])

Similar to Nietzsche's suggestion about humans transcending their current state as the path to become *der Übermensch* ('the superman'), transhumanism sees it as the path to reach the *posthuman* (Sorgner, 2009), a being with vastly greater capacities than present humans. This is not to say that Nietzsche's concept of the superman corresponds to the posthuman concept promoted by transhumanists. Certainly Nietzsche's idea of overcoming our current condition has more similarities with other conceptions of posthumanity than those held by most transhumanists. For example, it is unlikely that technological transformation was what Nietzsche had in mind as a path to transcend our current human condition; he was probably thinking more along the lines of elevated personal growth, cultural refinement and liberated aesthetic sensibilities achieved through philosophy (Bainbridge, 2010). In contrast Reade and Wells, faced with what they saw as fundamental changes driven by a new scientific outlook in the second half of the nineteenth century, created awe-inspiring visions of a future based in a teleological notion of techno-scientific progress (Coenen, 2014). These early transhumanist visions were further developed by a number of important scientists, such as J. B. S. Haldane, Julian Huxley and John Desmond Bernal, all of whom added to them a higher degree of technoscientific imagination (Coenen, 2014).

It was Julian Huxley, a distinguished biologist and Aldous Huxley's brother, who first used the term 'transhumanism':

> The human species can, if it wishes, transcend itself—not just sporadically, an individual here in one way, an individual there in another way—but in its entirety, as humanity. We need a name for this new belief. Perhaps *transhumanism* will serve: man remaining man, but transcending himself, by realizing new possibilities of and for his human nature.
>
> (Huxley, 1957)

Following Huxley's quote, the 'trans' prefix in transhumanism conveys, then, the idea of transformation or transcendence of the present human condition. While Huxley's view on transhumanism suggests that we could steer the direction of our evolution, he actually stressed the view of 'man remaining man'; thus he was not necessarily pointing towards a new nature in the sense that contemporary views on transhumanism suggest. Huxley's view, compared to that of Bernal or Haldane, was not focused in using technology to attain transcendence. It took a few years before transhumanism, as a cultural, social and philosophical movement, started to be focused on the use of technology to shape the values and lifestyle of current humans, and thus constitute a *trans*itional stage between humans and posthumans (Birnbacher, 2009; Bostrom, 2005a, 2008, 2009; Fukuyama, 2002; Gray, 2000; Miah, 2003, 2009; Pepperell, 2004, 2005). In this regard, cyberculture—a culture based on and mediated by the use of information and communication technologies (ICTs)—has greatly influenced that particular view on transhumanism.

In 1998 philosopher Nick Bostrom together with David Pearce took transhumanism to a different level when they decided to create the World Transhumanist Association (WTA).[28] Today we can find a rapidly expanding range of transhumanist groups engaged in a variety of discussion groups around the world, differing somewhat in focus and scope—such as extropians, singulitarians and academic transhumanism.[29] For instance, transhumanism has been used as a source of inspiration in artistic creation and cultural activities. Natasha Vita-More and Sterlac are prominent figures in mixing transhumanist ideas and art.

Today transhumanism seems to be torn between its Enlightenment conviction of inevitable progress towards the posthuman condition and new technological horizons, such as the singularity, and its rational awareness of the possibility that in the path towards progress more and newer risks to humanity may come along (Bostrom, 2002; Hughes, 2010b). Even though some transhumanists are starting to focus on reducing the potential risks of new technologies and the responsible use of science and enhancement technologies, the view of transhumanism under this paradigm is one focused mostly on an optimistic view of the outcomes of technological change and innovation, while playing down the risks and the negative scenarios that might come along.

Transhumanism: The path to a posthuman era

Albert Camus wrote that 'man is the only creature that refuses to be what he is' (Camus, 1954, p. 11), an idea that seems to underlie the transhumanist goal to transcend the human condition. According to transhumanists the current state of the human 'need not be the endpoint of evolution' (Bostrom, 2005c, c.f. 2005a; Cascio, 2006; FM-2030, 1989; Garreau, 2005; Harris, 2007; Kurzweil, 2005). They view the human as a work in progress with many inherent biological limitations. According to transhumanists these limitations have only made accessible to human experience a small fraction of what is possible. Bostrom, for instance, has argued that 'in addition to being at the mercy of a genetically determined set point for our levels of well-being, we are limited in regard to energy, will-power, and ability to shape our own character in accordance with our ideals' (Bostrom, 2005c). Thus, it is assumed that there are other ways of living, thinking, relating and feeling that are not yet accessible to humans but which are valuable and desirable for one's own personal well-being and for the development of the species as a whole. Accordingly, transhumanists hold a view in which science and technology is the path that would eventually enable humans to reach a posthuman condition: a condition in which more and new modes of being could be experienced, including to 'learn more, see more, experience more, and understand more, including even more radical ways to change ourselves to seek truth and pursue the good' (Hopkins, 2008, p. 5).

The core idea underlying transhumanism is that technology will help humans transcend their all-too-human limitations. Even when some transhumanists, like Bostrom, have suggested that transhumanist ambitions go beyond technological gadgets to social change, economics, institutional designs and cultural development, the main trend and the one that this paradigm aims to capture is the technological type of transhumanism. Considering this, it is not surprising that transhumanist ideals have been predominantly common among individuals and nations with access to the types of technologies needed to achieve transhumanist ideals. Indeed, most transhumanist thinkers are Western individuals living in high-technological hubs or environments. Interestingly, as Bainbridge has pointed out (2010), with a few notable exceptions, leading transhumanists are not engineers or scientists, but philosophers, ethicists, sociologists and artists.

Transhumanist ideas are gaining momentum in the mainstream as a variety of journalistic works have focused on promoting transhumanists ideas, such as the magazines *Humanity+*, *Wired* and *Third Culture*. It has also shaped and is being shaped by academic bioethical discourse. A clear example of this is the paper entitled 'Towards responsible use of cognitive-enhancing drugs by the healthy', co-authored by several scientists and academics (Greely et al., 2008) in the journal *Nature*. While the authors of the paper thought it was an open-minded but cautious approach (Greely, 2010), it ended up being quite controversial. For example, some scholars regarded this paper as clear evidence of the influence that transhumanist ideas can have in the social and political context (Coenen et al., 2009), arguing that *Nature* 'obviously supports the far-reaching transhumanist visions of human enhancement' (Coenen et al., 2009, p. 100). The flagship publication of the Institute of Electrical and Electronics Engineers (IEEE), *Spectrum*, another well-known publication for scientists and engineers, also released a special issue devoted to the idea of 'the singularity'.[30] In this, different definitions of the technological singularity, a point in time at which aided by technology we would have created or become creatures of more than human intelligence, are touched upon as well as various views from those who consider the notion likely and those who think it is totally bogus. Furthermore, recent technological advances are blurring the line between science fiction and science facts, and might already have the potential to bring to light the kind of radical changes envisaged by transhumanists (Bioethics, 2012; Farah, 2004; Garreau, 2005; Joy, 2000; Roco & Bainbridge, 2002; Sandler, 2014).

These examples are evidence of how in a relatively short period of time transhumanism has started to gain a place in the public arena, and give us reasons to keep a critical eye on its developments in the coming decades.

Understandings of posthumanism

Throughout humankind's history attempts have been made to grasp what it means to be human from many diverse perspectives, including philosophy, psychology, anthropology, biology and history. With our increasingly sophisticated technological developments this historical attempt is once more being challenged as notions that have been used to define the human are being contested and blurred even more. In this context, the posthuman has become a very important concept not only within the human enhancement discourse, but also in contemporary literary theory, the sociology of the body, philosophy, cultural and film studies, and art studies. The vision that transhumanists have of posthumanity is then just one of the different—and possibly conflicting—understandings of posthumanity, as the concept involves a range of cultural, political and technical elements (Cabrera, 2009b; Hayles, 1999; Miah, 2009; Sorgner, 2009).

Posthumanism is an umbrella term that has been used to understand the displacement of the human, humanism and the humanities (Halberstam & Livingston, 1995; Hayles, 1999). For example, Michel Foucault's suggestion in *The Order of Things* that man is a historical construction whose era is about to end (1970) can be regarded as an account of posthumanism. Ihab Hassan, an Egyptian literary theorist, has also widely written about postmodern culture:

> We need first to understand that the human form—including human desire and all its external representations—may be changing radically, and thus must be re-visioned ... five hundred years of humanism may be coming to an end as humanism transform itself into something we must helplessly call posthumanism.
>
> (1977, p. 212)

The kind of posthumanism implied in these accounts does not seem to have anything to do with the technologically enhanced human vision suggested by transhumanists. Even taking a technologically based vision of the posthuman, there are various perspectives that one can take. For example, posthumanism of disembodiment wants liberation from the limitations of the physical realm, whereas posthumanism of embodiment recognises continuities between realms that might be considered as distinct and bounded (Pepperell, 2005). The following section will explore two main visions of posthumanity, namely philosophical posthumanism and cultural posthumanism (Miah, 2009).

Philosophical posthumanism

Philosophical posthumanism is engaged in continuing the Enlightenment ideal that progress can be brought through the employment of technology. Others have regarded this view of posthumanism as 'speculative posthumanism' (Roden, 2010) or as transcendent posthumanism

(Munkittrick, 2010). This view of posthumanism embraces the use of new technological means as a way to help us redesign nature and ourselves 'into varieties of intelligent life' (Hughes, 2004, p. 140), and to enable us to explore different 'modes of being' (Bostrom, 2009). Taking into account the key features of philosophical posthumanism, it is not hard to see that transhumanist ideas of posthumanity followed this view.

Under this understanding of posthumanism, a standard view on the posthuman is the one suggested by Bostrom, who defines the posthuman as a being who has attained at least one posthuman capacity after extreme human enhancement (Bostrom, 2009); a posthuman capacity being a 'general central capacity'—intellectual, physical or psychological—that greatly exceeds 'the maximum attainable by any current human being without recourse to new technological means' (Bostrom, 2009, p. 8).[31] This view of the posthuman emphasises the use of new technological means to enhance individuals in order to attain posthuman capacities.

Supporters of this view acknowledge that although not all posthuman capacities would be worthwhile attaining, it could still be good to become posthuman and experience different modes of being. However, not everyone agrees that posthuman capacities are appealing or that they actually bring any value. For example, Francis Fukuyama in his book *Our Posthuman Future* (2002) described the posthuman as the result of the immorality of human enhancement. According to Fukuyama the posthuman is a biotechnologically transformed creature that has a human shell but is already separated from its natural biological origins, a successor of what would become an obsolete biological human.[32] Hence, according to this view the posthuman is the result of having lost key features of our humanity through technological interventions. Nonetheless, as Mark Smith (2005) has stated, a more open debate is required to clarify whether those who advocate technological development towards 'the posthuman' differ or not from those who simply advocate 'radical' technological solutions to human physiological, intellectual and psychological limitations and impairments.

Cultural posthumanism

A completely different concept of the posthuman is the one suggested in cultural posthumanism, which is also regarded as critical posthumanism (Roden, 2010). This view of posthumanism draws on post-structuralism, deconstruction and theories of embodied cognition; thus it focuses on a narrative of *otherness* and the capacity to be politically divisive (such as destabilisation of humanist values). It also involves a series of interactions between evolving/devolving incorporations and inscriptions, and a suggestion that our concept of the human as a natural, non-technological being was probably mistaken from the beginning.

A standard view of the posthuman under cultural posthumanism is the one suggested by Katherine Hayles in her book *How Do We Became Posthumans* (1999). According to Hayles the posthuman is 'an amalgam, a collection of heterogeneous components, a material-informational entity whose boundaries undergo continuous construction and reconstruction' (Hayles, 1999, p. 3). Thus, for Hayles the posthuman subject is not necessarily embodied in a material substratum, but rather is deconstructed and disembodied as information. Another definition of the posthuman within the cultural posthumanist is the one given by Donna Haraway, for whom the posthuman—or as she prefers to refer to it 'the cyborg'—is a political tool to reconstitute our political identity (Haraway, 1991). Manfred Clynes and Nathan Kline in the 1960s coined the term 'cyborg' to refer not only to humans but to any living creature, whose capabilities, bodies and features have been extended with machines (Clynes & Kline, 1960). Strictly speaking, according to this definition, anyone with an artificial organ, limb or supplement, anyone altered in any technological way to think, behave, feel or look different can be regarded as a cyborg (Clark, 2004; Gray, 1995). Cyborgisation is only a possible route to posthumanity, which neither guarantees the realisation of the techno-posthuman nor the involvement of enhancement (as understood in the transhumanist view). That is why most people supporting a cultural view on posthumanism seem to prefer to use the term 'cyborg' rather than transhuman.

The posthuman envisioned by cultural posthumanism, contrary to philosophical posthumanism, does not need to be a techno-enhanced being with newly attained modes of being. As Hayles has pointed out, the posthuman is more about the 'construction of subjectivities' rather than the 'presence of nonbiological components' (1999, p. 4). In that regard even a biologically unaltered human can count as a posthuman. For example, individuals immersed in virtual reality or using constantly online social networks are examples of this type of posthuman, insofar as these activities enable the possibility of different subjectivities and identities. Under this view, even people who through their art or meditation practices experience different subjectivities can be counted as posthumans. This does not mean that cultural posthumanism is not concerned with technology; rather, it has positioned technology in relation to the differences observed and developed theories of change

(Miah, 2009). Furthermore, it regards technology not so much as a tool to *attain* new modes of being, but rather as a process that *reveals* specific modes of being (Heidegger, 1982).

Even though philosophical and cultural posthumanism might hold different conceptions about the posthuman, both accounts seem to agree with the following ideas: the posthuman (1) implies a new horizon relatively inaccessible to untransformed humans, in which posthumans acquire, albeit by different means, new capacities to act and to be affected; (2) involves a redistribution and construction of difference and identity; and (3) is an attempt to revise different views about what it means to be human, within the context of radical changes, whether instantiated by emergent technologies or deconstruction of meaning.

Now that we have a better idea of the foundational concepts underlying the transhumanist human enhancement paradigm, the next section will develop the main features of the transhumanist paradigm.

The transhumanist paradigm: Towards posthumanity

Within the biopolitical spectrum of human enhancement,[33] the transhumanist paradigm can be regarded as the opposite of the biomedical one. While the biomedical paradigm holds a bioconservative view regarding human enhancement, the transhumanist paradigm holds a more liberal view. Furthermore, transhumanists can be considered as a subgroup of the biopolitical groups sharing a commitment for technological advancement, such as bioliberals (Roache & Clarke, 2009) or technoprogressists (Hughes, 2010a).

The transhumanist paradigm is still a small fraction of the different human enhancement views. However, as mentioned above, given the fact that the transhumanist movement is gaining attention and support around the world, there are reasons to think that it could become increasingly popular, strongly influencing people's perception of enhancement. This is more likely to be the case among high-technological societies that regard highly autonomy and freedom, and that hold a liberal view on the individual.

Transhumanist enhancement

Transhumanist enhancements under this paradigm are mostly focused on radical technological interventions aimed at enhancing human capacities. While biomedical enhancement was about interventions beyond therapy, but still within the limits of species-typical capacities, the transhumanist paradigm is focused on the attainment of posthuman capacities, that is enhancement of capacities *beyond* species limits (Bostrom, 2005a, 2005c, 2008; Garreau, 2005; Miah, 2003; Naam, 2005; Pepperell, 2005). As a result of this, transhumanist enhancement interventions have been regarded as 'second-stage enhancements' (Khushf, 2005).

Examples of transhumanist enhancement views include 'repair that makes the individual more capable than they were prior to the need for repair, and more significantly, to a level that exceeds the capabilities of all human beings' (Miah, 2003, p. 5), or interventions that would allow us to go beyond our current biological limitations, enabling us to explore a 'larger space of possible modes of being' (Bostrom, 2005a, 2009), which otherwise we have not been able to explore. Considering these different views, transhumanist enhancement can be regarded as:

> **Transhumanist enhancement:** any intervention, not necessarily medical, aimed primarily at the improvement of one or more core capacities of an individual beyond *species-typical limits* with the aim of *overcoming* human biological limitations.

The paradigm

This paradigm retains an individual-based focus but it is aimed at beyond *species-typical* boundaries; thus it can be regarded as a transhumanised version of the biomedical paradigm (Wolbring, 2008a). The transhumanist paradigm, however, takes a different moral view of enhancement. Whereas the biomedical paradigm assumes that therapy and enhancement are morally different—therapy being morally permissible while enhancement is not—this paradigm does not consider them as morally different practices.

The main reason offered for holding such a view is that individuals seek enhancement for the same reasons they seek therapy; namely, to improve their lives—by attaining better health, emotional well-being or protection from harms. Thus if therapy is usually not regarded as morally dubious, enhancement cannot be regarded as morally dubious (Bostrom & Roache, 2007; Harris, 2009; Naam, 2005). This paradigm considers human enhancement, then, not as intrinsically wrong, ethically controversial or particularly risky, but to the contrary as valuable for humans. Instances of the value that enhancement can bring, according to supporters of this paradigm, include radical extension of human lifespan, elimination of suffering, eradication of disease and augmentation of human intellectual, physical and emotional capacities (Bostrom, 2005c, 2008; Bostrom & Roache 2011; Kurzweil, 2005). Some even think human enhancement should generally be permitted and promoted, or even considered a moral obligation (Harris, 2007, 2009).

This paradigm, as well as the biomedical paradigm, is based on and promotes individualistic and individual-based values. Even in cases where social benefits—such as revolutionary changes in health care, improvement of group creativity and independent thinking, highly effective communication techniques, extended altruism and self-control, or a strong sense of justice and fairness—are brought up, these are seen rather as the result of cumulative individualistic interventions. This paradigm, in contrast with the biomedical one, takes a more liberal view of the right that individuals have to choose which enhancement interventions, if any, they want (or would like) to undertake. Its emphasis on *individual freedom* and *individual choice* is visible in their arguments defending morphological freedom—the individual right to transform one's own body as desired (Sandberg, 2003)—and reproductive freedom. Thus, within this paradigm coercive interference with individual freedom and choice or the mere fact that others might feel disgusted or morally disrespected by a particular kind of enhancement are not considered as legitimate grounds to prevent people from pursuing enhancement. It is worth noting that even though some transhumanists have argued that some restrictions in individual enhancement options are legitimate 'where individual choices impact substantially on other people' (Bostrom, 2005c), this is not the general view.

The transhumanist paradigm of human enhancement, similar to the biomedical paradigm, is focused on enhancement interventions that directly change individuals' capacities. However, it also pursues a broader dimension—the improvement of the *species*. This dimension becomes clearer when we think about transhumanists' ultimate goal of enhancement, namely the attainment of posthumanity, a future society integrated by technological-based posthumans.

While posthumans are regarded as beings entitled to moral status (Douglas, 2011), supporters of this view of enhancement have argued that this does not necessarily entail a higher moral status than that attributed to unenhanced humans (Bostrom, 2005b, 2007; Hughes, 2004). A related feature of this enhancement paradigm is its perspective on the nature and development of beings worth of moral respect. According to transhumanism it is persons who are worth of moral respect. An approach based on personhood acknowledges that not all human qualities are morally relevant and that some non-humans share these relevant qualities. Among the relevant features that have been suggested for granting personhood are: awareness of oneself across time and space, being capable of feelings and emotions, having a sense of self and consciousness, being able to recognise other persons and treat them accordingly, showing a variety of cognitive abilities (such as analytical and conceptual thought, intelligence or different degrees of communication). It is in virtue of possessing those relevant qualities that all persons, including posthumans if they have the relevant features, should be treated with equal moral concern (Farah & Heberlein, 2007; Savulescu, 2009). The extreme version of this line of argumentation leads to the view that it is morally permissible to kill human beings who are not persons (such as foetuses, victims of deep coma or individuals with certain cognitive impairments), but morally wrong to harm persons even if they are not human beings.

Other features of this enhancement paradigm include an open and embracing attitude towards new technology. Thus, people supporting this paradigm do not welcome approaches that try to ban or prohibit certain technologies. Furthermore, under this paradigm, whether enhancement interventions are inside or outside the body, or whether they are permanent or temporary, make no difference to the moral equation. Thus, using a brain implant for improving our neural connections is not more different, morally speaking, than taking vitamin supplements for the same purpose (Bostrom & Roache, 2007; Harris, 2009). There is no deep moral difference between technological and other means of enhancing human lives. Thus, supporters of this human enhancement paradigm charge those who think otherwise with being morally arbitrary.

The idea of treating different enhancement interventions (e.g. internal and external, or high technology and 'natural') with moral parity uses a similar standpoint to the one suggested in Andy Clark's and David Chalmers's *parity principle*. Clark and Chalmers introduced this principle in their paper 'The Extended Mind' (1998) by using the example of shifting an object (such as a Tetris figure) by three different means. In one case your biological 'unchanged brain' does the computation (that is to say, you mentally do the shifting of the object). In another case a computer makes the computation after you press a button that rotates the object on a screen. In the final case your 'enhanced' brain (for instance with a neuroimplant) performs the computation. Clark and Chalmers suggest that all three cases are similar, calling it the *parity principle*. According to this principle, the part of the world that functions as a process as we confront some task, which we would have no hesitation in recognising as a process that could have been carried out in the head (cognitive process) becomes (for the time we are engaged upon the task) part of the cognitive process.[34] Thus, in their example there is parity between the three cases.

Even though under the transhumanist paradigm moral parity between technologically based enhancements and other forms of enhancement is acknowledged, high-technologically based enhancement interventions are given preference. Consider the following statement:

> [t]here are limits to how much can be achieved by low-tech means such as education, philosophical contemplation, moral self-scrutiny and other such methods proposed by classical philosophers with perfectionist leanings... This is not to denigrate what we can do with the tools we have today. Yet ultimately, transhumanists hope to go further.
>
> (Bostrom, 2005c, p. 5)

Considering this feature of the transhumanist paradigm, the neuroimplant would be the preferred option to rotate the object in the Tetris game example mentioned above, because in theory it allows us to perform the computation faster than using mental rotation and faster than having to press the button.[35] The focus seems to be more on attaining high performance and efficiency than on the possible risks and side effects involved. Given that this paradigm supports predominantly the use of radical technological interventions for enhancement, we can see the key role that emergent technologies are given within the transhumanist agenda.

The role of emergent technologies in the transhumanist paradigm

Emergent technologies have been regarded as holding the potential to change the human condition in radical ways (Cabrera, 2009a; Milburn, 2002). This vision is at the core of this human enhancement paradigm. In most reports connected to new technologies one can find a transhumanist touch. This becomes evident, for instance, when it is stated that the convergence of these technologies will enable us to deal with human challenges by 'substantially enhancing human mental, physical and social abilities' (Roco & Bainbridge, 2002), or that the convergence of these technologies has huge potential to alleviate human suffering and to accelerate access to sustainable energy, abundant food and universal healthcare (Canton, 2004; Roco & Bainbridge, 2002; Zonneveld, 2008). Nonetheless, that is not the real novelty of the transhumanist agenda, but rather the idea of technologically reshaping or redesigning ourselves in radical, ubiquitous and permanent ways (Coenen et al., 2009; Gordijn, 2006).

The use of technology under this paradigm is focused on redesigning the human condition by offering *individual* interventions to overcome *species-typical* biological limitations. The forms of enhancement envisioned by this paradigm are somehow more speculative and radical than those in the biomedical paradigm, as they are aimed at attaining features *beyond* those typical of the species, also referred to as posthuman capacities. The majority of these interventions, as in the case of the biomedical enhancement paradigm, are individual-focused, individualistic and primarily aimed at directly changing the individual.

There are two main premises underlying the use of emergent technologies for human enhancement under this paradigm. The first is that we are no more than sophisticated machines made up of billions of biomolecules interacting according to 'well-defined, though not completely known, rules deriving from physics and chemistry' (Brooks, 2008, p. 1). The second one is that the human brain and body can continuously be improved—just as another one of our machines might be.

Some of the specific technological interventions discussed under this paradigm include the use of emergent technologies for the creation of heterogeneous molecular nanosystems. The idea behind these systems is that each molecule will be used as a device with specific structure and role in order to enable new functions in the system and even to enable radical extension of human lifespan. In this regard, nanomedicine is considered to be a great ally for achieving these goals (Freitas, 1999, 2005b, 2007). Evolutionary cells, self-replication of large nanostructured systems and neuromorphic engineering are some examples of molecular nanosystems envisioned under this paradigm. Other interventions have to do with directly intervening in our brain. A clear example of this idea is contained in the following statement: '[n]eurons could be re-engineered so that our minds "talk" directly to computers or to artificial limbs' (ETC Group, 2005, p. 8). According to some researchers, it is non-biomedical forms of enhancement that will most likely instantiate the most dramatic advances (Bostrom & Sandberg, 2009a; Coenen et al., 2009; Persson & Savulescu, 2008).

Neural interfaces and nanobots are among the emergent technologies applications that have captured more attention, most likely as they have the potential to bring about the sort of posthuman features envisioned by the transhumanist paradigm. Another reason for their having captured so much attention is that the military and entertainment sectors have great interest in their possible applications. The US Defense Advanced Research Projects Agency (DARPA), for instance, is involved in many projects that are aimed at improving and creating new brain machine interfaces, such as 'thought helmets' that could in the future secure mind-to-mind communication between soldiers, or neural interfaces in which military personnel can be wired to silicon circuits of the equipment they control. In the entertainment sector, similar technologies could enable video game players to use their brainwaves to control features of the game with more precision, for instance, by reading emotions of players that can be translated into the virtual world.

Other envisioned interventions are neuroimplants that could enable people to learn things without going through the actual experience (i.e. making a complex evasive manoeuvre on a plane), or could enable brain-to-brain communication (Warwick, 2004, 2014). A related transhumanist enhancement is mind-uploading. The general idea is that human memories can be extracted (uploaded) from the brain and imported, intact and unchanged, to digital format either in order to inhabit a robot body or to allow a person to live in virtual reality or cyberspace.[36] Some people have suggested that this could be achieved by using nanobots that would scan every corner of our brain or alternatively by reverse-engineering our brain. In either case the idea is to disassemble our protein-based brain and assemble it again in a silicon-based brain in order to obtain the pattern of our memories and other information that we have stored, which would later be processed by computational models and supercomputers (Bostrom, 2005a; Drexler, 1986; Kurzweil, 2005). The processed information could be used to simulate certain human capacities and eventually to emulate the whole computational structure of our original mind (Sandberg & Bostrom, 2008), to store our memories in digital format, and perhaps even to edit, erase and add new memories (Bostrom, 2005a; Kurzweil, 2005).

Most transhumanists envision that mind-uploading will enable us to reach immortality. They believe that insofar as our neural patterns are the essential aspect of who we are, we could upload our memories pattern to a computer or any other technological device, and live in cyberspace even after our body has died. Some transhumanists have gone even further, arguing that these uploaded experiences and memories would be able to link-up in a global network, enhanced individuals forming a kind of hive mind. Under the transhumanist paradigm the idea of sharing memories as suggested by this application and brain-to-brain communication implants is considered an enhancement as it would enable better understanding of other people's points of view as well as the exploration of other people's (or other beings') perceptions of the world.

Enhancement interventions under this paradigm are seen as the path to explore different modes of being and attain posthuman capacities. Thus, the idea of merging with our technologies is not feared; rather it is cherished and promoted. Moreover, it is assumed that humans could benefit from the different ways in which our technology can be easily networked, have more and longer-term memory capacity storage, and process information in hundreds of dimensions and faster (Warwick, 2004). All of these desired features come mainly from the features of ICTs. This of course stresses the assumption held in this paradigm that humans are reduced to machine-like beings.

In the last section of this chapter, the main arguments used by transhumanists to support enhancement will be analysed, followed by some of the main problems the paradigm faces and finally some valuable ideas that we can learn from it.

The ethics of transhumanist human enhancement

This section analyses some of the most recurrent arguments used by transhumanists to support human enhancement. The arguments presented here follow a similar order from those discussed under the biomedical paradigm, but here, contrary to the anti-enhancement perspective of bioconservative arguments, these hold a pro-enhancement perspective. It is important to keep in mind that the core value of transhumanism is to explore the different modes of being that the transhuman and the posthuman enable. Other values underlying transhumanism are individual choice in the use of enhancement technologies, morphological freedom and diversity, life extension and the idea of hubris rejected (Bostrom, 2005c).

The arguments

Naturalness

Appealing to the naturalness argument is not enough to refute human enhancement. First of all, there are natural things that we do not necessarily associate with virtue or being good for us, such as disease and natural disasters. Second, the natural versus artificial distinction is not an easy one to sustain given that we can use the term 'natural' in many ways. Natural can be used to refer to following the fundamental laws of nature. If humans are regarded as part of nature, then the changes that we make to nature or ourselves cannot be considered as less natural.

For instance, we generally do not consider the changes beavers make to their environment through dam-building to be unnatural (Bayertz, 2003). Following this line of argument, everything humans do is natural, including our use and development of technology for human enhancement. In addition, if the desire for enhancement is intrinsically human, then it cannot be regarded as morally wrong (Bayertz, 2003).

But natural can also mean existing without human intervention. Under such an interpretation the implication would be that everything we do is unnatural. Thus, if we based our moral distinctions on this understanding of nature we would end up with scenarios in which eating canned food, reading books and using computers would be morally reprehensible. Moreover, if the idea is that we should not use our technologies to improve ourselves because it is not natural, then we can think that humans 'have never been "natural"' (Rose, 2007, p. 80). Third, moral views about nature tend to rest on feelings and fears about the prospect of change; thus it is most likely that they are biased and not well grounded in rational principles (Bostrom & Ord, 2006; Levin, 2003). As Haldane wrote:

> There is no great invention, from fire to flying, which has not been hailed as an insult to some god. But if every physical and chemical invention is a blasphemy, every biological invention is a perversion. There is hardly one which, on first being brought to the notice of an observer from any nation which has not previously heard of their existence, would not appear to him as indecent and unnatural.
>
> (Haldane, 1923)

Finally, under this paradigm interventions in nature are not considered as a way to challenge God or mess with the divine creation, but rather, can be regarded as being consistent with religious views. In this context Eric Parens has argued that enhancement can be considered as part of our responsibility 'to use our creativity to mend and transform ourselves and the world' (Parens, 2006, p. 76).

Considering the previous points, it might be concluded that morality cannot be fundamentally about countering nature, or good be associated with what is natural or bad with what is unnatural. Thus, naturalness is not a good starting point to discuss the normative implications of human enhancement.

Autonomy and freedom

Interestingly enough, autonomy and freedom are not only used in arguments against human enhancement but are just as often offered as defences of it (Harris, 2009). Contrary to the biomedical paradigm which considers enhancement as a threat to autonomy and freedom—a fear which according to transhumanists is based on an outmoded Kantian metaphysics of free will and moral absolutism (Smith, 2005)—the transhumanist paradigm sees human enhancement as a key manifestation of our autonomy and freedom. On the one hand, enhancement is an affirmation of our capacity as individuals to freely acquire or develop values or interests and act accordingly. On the other hand, people's right to alter their bodies (morphological freedom), genes (genomic freedom), intellect (noological freedom) and mental functioning (psychological freedom) can be regarded as an extension of our basic rights, such as the right to life, to liberty or to security. That is why any restriction on enhancement would infringe upon our fundamental ability to choose how to live our lives (Bostrom & Savulescu, 2009; Bublitz & Merkel, 2009; Harris, 2007; Naam, 2005).

Human dignity

Features that were considered the foundation of human dignity, such as consciousness or intelligence, have turned out to be not as uniquely human as we used to think. Some scholars even consider the concept to be 'useless' because it is (a) a vague restatement of other more precise notions; (b) falls within a group of notions about capabilities and social interactions; (c) can be reduced to autonomy; and (d) is based on 'human prejudice' (Bostrom, 2007; Jacobson, 2012). Transhumanists have focused on the latter point to counterargue bioconservatives' arguments. In particular, they argue that by invoking a threat to human dignity bioconservatives are using an unsound and outmoded argument that defends anthropocentrism, one based on abstract and potentially fictional identifiers (Smith, 2005).

Transhumanists disagree with bioconservatives in that posthumanity need not be a threat to dignity; rather they consider human enhancement interventions to have the potential to increase dignity (Bostrom, 2007). Bostrom, for instance, regards dignity as 'a quality, a kind of excellence admitting of degrees and applicable to entities both within and without the human realm' (Bostrom, 2007, p. 173). Therefore, posthuman dignity and human dignity can be regarded as compatible and complementary (Bostrom, 2005b; Hughes, 2004). By defending posthuman dignity, transhumanists see the promotion of 'a more inclusive and humane ethics, one that will embrace future technologically modified people as well as humans of the contemporary kind' (Bostrom, 2005b, p. 213). Similarly, Hughes (2004) has argued that once we have enhanced ourselves it might be better to talk about dignity based on personhood rather than on being human. Under such a view there would not be a conflict with human dignity since we would grant this 'respect' to all persons.

Social disruption

Most of the social disruption issues charged against enhancement interventions by supporters of the biomedical paradigm are considered under the transhumanist paradigm as lacking evidence. However, there are two issues that at least some supporters of this paradigm do acknowledge, dual use and existential risks. The latter deals with those risks where 'an adverse outcome would either annihilate Earth-originating intelligent life or permanently and drastically curtail its potential' (Bostrom, 2002, p. 2); whereas the former deals with cases such as terrorists using certain enhancement applications for bad purposes.

In contrast with the pessimistic view around the social impact of enhancement interventions held by the biomedical paradigm, the transhumanist paradigm is overly optimistic about the benefits that human enhancement could bring, in particular for individuals. This overly optimistic view is based on two main premises. The first is that the benefits of human enhancement exceed any possible negative issues. The second is that human enhancement would help us fix initial problems that might come along. Bostrom and Ord (2006), for instance, claim that despite the fact that certain enhancements might bring forward novel risks, they might also bring along ways to reduce other fatal threats to humanity.

Authenticity

As mentioned in the previous chapter, there are two interrelated ideas underlying the authenticity argument, cheating and the good life. In the case of cheating, most people would agree that natural endowments are not equally distributed. However, we would generally not regard as cheating instances where athletes are born with a better aptitude than the rest of us for running. Interestingly, people's intuitions are more diverse when considering the use of advanced prostheses by amputee athletes competing against able-bodied athletes. In the past, when prostheses were not very good, amputees were at a disadvantage compared to able-bodied athletes and as such it was not fair for the amputee to compete against them. However, with more sophisticated prostheses things are turning the other way around. New prostheses are becoming so advanced that they defy the limits that able-bodied runners have previously set. This has triggered a debate about whether it is fair or not for able-bodied athletes to run against amputees using this type of prosthesis. From this, we can say that views around excellence and accomplishment are measured relative to standards that are dynamic (Allhoff, 2005). Therefore, enhancement should not be morally controversial just because it would augment performance or require less effort from the individual. In connection with this Bostrom has argued that 'the possession of posthuman capacities could be extremely valuable even where the capacities are effortlessly obtained' (Bostrom, 2009, p. 22). The fact that they are effortless does not make them necessarily a form of cheating; that would depend on the standards set by society and the conditions under which enhanced individuals will be competing.

Connected to the idea of the good life, the general argument provided by this paradigm of enhancement is that people search for human enhancement as a way to improve their well-being and contribute to their attainment of the good life. If enhancement helps us in the pursuit of excellence, how can it be regarded as morally dubious? According to the transhumanist paradigm, human enhancement interventions are just another path towards the good life insofar as they enable us to attain posthuman capabilities.

Evolution

This last argument is a reply to the playing God argument that is discussed in the biomedical paradigm. According to this paradigm, human enhancement is not a case of humans playing God, rather it is merely the next logical step in the ongoing process of human evolution. Therefore, we can refer to this as the evolution argument. Under this paradigm the human condition is regarded as not optimal, thus enhancement is the path that takes humankind into a *better* condition. Consider Kurzweil's answer to someone arguing against enhancement: *If you're speaking for yourself, that's fine with me. But if you stay biological and don't reprogram your genes, you won't be around for very long to influence the debate* (Kurzweil, 2005, p. 226 [emphasis in original]).

Bostrom and Sandberg (2009b) propose the evolutionary optimality challenge (EOC), as a way of analysing why we have not yet evolved into the type of beings that transhumanist enhancement interventions will enable us to become. They argue that nature has not *given* us certain traits, because it is (a) fundamentally incapable of producing them, (b) does not count on the tools and knowledge that we have to shape our own evolution and (c) simply because it takes too long to develop those traits. The second line of argument they use is that evolution has not optimised us for certain traits that would be preferred at a global scale, such as happiness (Bostrom & Sandberg, 2009b). Bostrom and Ord made a similar point in their paper about status quo bias by stating that 'there is no general reason for thinking that what evolution selects for us—inclusive fitness—coincides with what makes our lives go well individually, much less collectively' (2006, p. 666).

Having rehearsed the arguments representative of this paradigm's normative stand, in the following section the main challenges this paradigm faces are explored.

The issues

The following are some of the main issues that this paradigm is confronted with. We will start by mentioning those that are similar to the biomedical paradigm and then move on to those that are particular to this paradigm.[37]

Shared issues with the biomedical paradigm

Both paradigms are mainly focused on the liberal individual, a view of the individual as abstract and isolated agent. Supporting this view, Hayles has pointed out that 'transhumanist rhetoric concentrates on individual transcendence; at transhumanist websites, articles, and books, there is a conspicuous absence of considering socioeconomic dynamics beyond the individual' (Hayles, 2010). A focus on the liberal individual, together with an emphasis on the possible benefits of individualised enhancement, not only puts an emphasis on competitiveness rather than on cooperation; it also highlights the unclear perspective we have about the value we attach to human life.

Furthermore, the idea of the liberal individual confuses the notion of the individual's freedom to 'choose' with the individual's freedom to 'obtain' (Sartre, 2012). Just as we saw with the biomedical paradigm, it is not clear why interventions that bring enormous benefits to *one* particular individual should be preferred to interventions that bring smaller benefits to *many* individuals. Daniels (2008) has, for instance, used the example of saving as many legs as possible (which presumably would improve or at least maintain the range of opportunities open to those individuals) rather than saving one single life. In the particular case of the transhumanist paradigm, a further challenge is to justify why we ought to be concerned with interventions that take the individual beyond the realm of human experience.

Another shared issue with the biomedical paradigm is the promotion of medicalisation. In the transhumanist case, however, medicalisation is not only about promoting more medical interventions but interventions that will take the individual beyond *species-typical* features. A final shared issue between these two human enhancement paradigms is the legacy of dualistic values regarding the body and mind. On the one hand, the healthy mind is associated with being quick, active, powerful and immortal. It stands for the domain of intelligence, the rational, responsibility and freedom. On the other hand, the body is associated with the opposite qualities: it is passive, weak and mortal; ultimately not governed by human will, but by the inexorable rules of the material world (Kirmayer, 1988). In the case of the transhumanist paradigm, it is not an exaggeration to say that these views are taken to an extreme, as implied by their suggestion that humans need to transcend their biological nature.

Technology mechanistic account

One main problem with this paradigm is its inherent technological determinism,[38] which refers to the idea that technology is the prime factor in shaping our lifestyle, values and society (Bijker, Hughes, Pinch, & Douglas, 2012; Heidegger, 1982; Kranzberg, 1986). Furthermore, the kind of language used (such as maximisation of efficiency, speed, precision, productivity) fosters a view in which we adjust human ends to match the features of our dominant technology, an idea that Langdon Winner calls 'reverse adaptation' (1977). Consequently, humanity figures more as an object to be improved than as a potential actor of change, where all our experiences and processes can be explained by science and technology, and where it becomes harder than ever to sustain the fundamental differences humans might have with their own technological creations (Fernández-Armesto, 2005).

Technological progress optimism

A related issue to this paradigm is blind optimism in technological progress (Coenen et al., 2009; Verdoux, 2009), or as others have called it 'techno-utopian inevitabilism' (Hughes, 2010b). While someone can be pessimistic about technology and its influence and still endorse transhumanism,[39] the transhumanist paradigm endorses a technoprogressive optimism view. Transhumanists' reliance on judgemental heuristics—principles or methods used to make assessments or judgements of probability simpler—has made them fall into different forms of cognitive biases (Tversky & Kahneman, 1974). For instance, by relying on representativeness as a judgemental heuristic, transhumanists tend to under/overestimate the evidence and over-interpret the findings, as well as underestimate the probability of failure in complex systems.

Over-generalisation and future-risk perception are the two other main cognitive biases related to technoprogressive optimism.

1. Over-generalisation argument:[40] the use of historical records to support transhumanist view on technological progress is on the one hand focused on the features of technology as a problem-solver rather than as a problem-generator (Gray, 2000; Kranzberg, 1986; Postman, 1998; Verdoux, 2009). Melvin Kranzberg's second law of technology, 'invention is the mother of necessity' (1986), is a good example of how we invent something in order to solve a problem, but then, as part of the trade-offs that come with such intervention, we are forced to invent something else to either solve a new problem or to make the previous solution work better.

On the other hand, transhumanists' general claims about technological progress are based on a highly impoverished selection of data extracted from a very particular period of history (Verdoux, 2009). Consider the following passage from Bostrom's 'Transhumanist Values':

> The history of economic and technological development, and the concomitant growth of civilization, is appropriately regarded with awe, as humanity's most glorious achievement. Thanks to the gradual accumulation of improvements over the past several thousand years, large portions of humanity have been freed from illiteracy, life-expectancies of twenty years, alarming infant-mortality rates, horrible diseases endured without palliatives, and periodic starvation and water shortages. (2005c)

Many of the improvements mentioned by Bostrom are improvements that have being brought forward by social and very low-technology interventions. Moreover, empirical data has also shown that the most violent, poor and vicious periods of human existence have resulted not from a lack of technology, as Bostrom's quote seems to imply, but rather from civilisation itself (including an excessive use of technology). Likewise, many of the conditions that enhancement technologies promise to halt and reverse—such as cancer, diabetes, heart disease and obesity—are mostly 'diseases of civilization' (Weder, 2007). It is hard to see the kinds of conditions that have come with advances in technology as progress, but similarly it would not be progress to return to a pre-technological state (Kurzweil, 2005; Verdoux, 2009).

2. Future-risk argument: an optimistic view on progress undermines the technogenic 'existential risks' brought forward by enhancement technologies.[41] In the light of enhancement technologies that can potentially instantiate such kinds of risks, transgenerational in scope and terminal in intensity, our understanding of progress might need to be revised.

Its vision of transhumanism and the posthuman

Another main problem with this paradigm is its particular vision about transhumanism and the posthuman, and the implications such views have for the human. Bioconservative supporters have argued that there are plenty of reasons to be worried about transhumanism. For instance, Fukuyama (2004) has argued that transhumanism is one of the most dangerous ideas ever as it threatens human dignity, which can lead to dehumanisation. Others believe that transhumanism is 'the ultimate illustration of how Enlightenment rationalism can easily run amok and create extreme pathology' (Ford, 2005). However, it is the hype and hope around the ideas it promotes, for example the blind faith in human perfectibility and technological progress, the lack of a broader discussion on its normative assumptions and the overrated appraisals of the state of the art made by some transhumanists that we should really be concerned about.

Most transhumanist writings present reasons why we should overcome our current human limitations, including a slow speed of calculation, distractibility and limited data-storage memory capacity, unreliable feelings (such as love), judgements that are easily clouded by emotions, an inability to perceive the world in different ways and the inability to live forever. The view that these are human limitations is not only questionable, but it also has interesting normative implications (Cabrera, 2009b; Coenen et al., 2009; Wolbring, 2008a). To start with, it implies that our current biological features are undesirable (and in some cases it further implies that they are impairments or disabilities) and therefore ought to be fixed or overcome. A couple of points can be raised as counter-arguments: first, these myriad 'shortcomings' are partly what differentiate us from the machines we created. Second, if we have not yet gained sufficient knowledge regarding the processes and features to be enhanced (such as a definition of morality or humanness), how can we have any reference point by which to measure the improvement? Furthermore, even if we agreed with the idea that such features are human limitations and that they are undesirable, we already have efficient ways to overcome most of them—computers, notebooks, data storage devices and special glasses. It can be further argued that the limits which we have not been able to overcome with the use of technology—such as the unreliability of our feelings and death—are part of being a complex life form.

The view of the human implied by this paradigm is one in which it is seen as intrinsically deficient or as a 'work in progress' (Wolbring, 2005), therefore in need of actualisation. The idea of the human as 'work in progress' is not problematic per se, as many existentialist thinkers have held similar views, in which the human is seen as a project in constant movement towards the future. Rather, what makes the transhumanist view problematic is that it views the human as just another cog in a wider technological project. This relates to the technology mechanistic issue discussed above, but it further stresses the reductionistic aspect brought forward by this paradigm.

The particular vision of transhumanism held by this paradigm has also promoted a transhumanisation of the concept of biomedical health. The transhumanist model of health sees every increase in functional ability or capabilities, even those beyond those who are typical of the species, as an increase in an individual's health. Superior ability itself is what is regarded as healthier. The consequence of such a view is that not only that people with impairments will be seen as *in need* of being treated in order to attain a more healthy state (as the biomedical model holds), but also that every single unenhanced individual will be seen not just unhealthy but perhaps even disabled (Cabrera, 2009b; Wolbring, 2008a). Here the term 'disabled' is understood as a 'diminished state of human being' (Campbell, 2001, p. 44), for lacking the 'normal' range of physical, intellectual, sensory or mental human features or capacities. Some people have even argued that those individuals who decide not to be enhanced are choosing to become a subspecies (Warwick, 2004).

According to this paradigm, the posthuman is the result of radical technology-based human enhancement that enables us to change our bodies and minds in new ways. Given the type of technology interventions envisioned for these changes, the view of the posthuman held by this paradigm is mostly oriented towards a small fraction of humanity, those with wealth, power and leisure to reconceptualise themselves (Hayles, 1999). In addition, this paradigm celebrates the posthuman as a higher mode of being (Bostrom, 2005a, 2009; Garreau, 2005; Pepperell,

2004, 2005) and puts it as the ultimate foundation of our world view. Bostrom's (2009) paper 'Why I want to be a posthuman when I grow up' is a good example of how the posthuman is seen under this paradigm. He argues that the posthuman implies the exploration of valuable modes of being that are not accessible to humans, that *some* possible posthuman modes of being would be very good, and concludes with the idea that the different modes of being enabled by the posthuman make it an 'exceedingly worthwhile type' of being for most current human beings (Bostrom, 2009, p. 4).[42]

While the exploration of posthuman modes of being might be interesting, it is not clear why *the posthuman* should be considered the main goal of enhancement as suggested by transhumanists (Bainbridge, 2010). In addition, even if the new modes of being enabled by the posthuman had value, this does not imply that there should be a common desire to become the kind of posthuman envisioned by transhumanists. Other arguments against this paradigm's view of the posthuman are based on premises such as: (1) the outcome of humans meeting posthumans would be very bad for human society; (2) posthumans will not hold the same values we attribute to current humans nor would they value the same things humans do; and (3) there is uncertainty regarding whether the posthuman will be an improved version of the human or a completely new entity. Connected to this latter point, we can ask ourselves how we can aspire to become *partially* human, *trans*human or *post*human when we are still debating what it means to be human.

The value of the transhumanist paradigm

Just as in the case of the biomedical paradigm, the transhumanist paradigm also has some ideas that are worth thinking through.

Morphological freedom: As a form of biological diversity

Humans have been changing and modifying their bodies for a long time—mostly for aesthetic reasons, by using cosmetics, ornamentation, tattooing and other culturally based body modifications. More recently, the disability movement together with postmodern critiques of the normativity of the human body have supported the idea of biological diversity (Miah, 2008; Smith & Morra, 2006; Wolbring, 2005). In a similar way to how people from the disability movement argue that individual differences can make an important contribution towards a biological diverse society (Miller, Parker, & Gillinson, 2004), transhumanists argue that the diversity obtained by human enhancement can have a similar positive contribution. Moreover, the paradigm considers self-modification as part of each individual's freedom. Anders Sandberg, for instance, sees self-modification, the idea of an individual freely deciding how best to alter his or her body as a basic right, and calls it morphological freedom (2003).

Morphological freedom within the transhumanist paradigm is mostly geared towards enabling individuals to attain aesthetic self-realisation, overcome the 'natural lottery' and as free experimentation with their bodies and minds. However, the concept can also be used to challenge long-standing views about our bodies. Power discourses have not only shaped what we regard as the norm, they have also promoted discrimination and intolerance towards those not fitting the established norm. Morphological freedom as a form of biological diversity (Miah, 2008) can help us question the enforcement of cultural norms about normality, to express what is truly humane in humans and add value to our lives. The case of people with impairments supports the idea that we can learn from diversity and variety (Cabrera, 2009b).

Diversity is not intrinsically a bad thing. A biological diverse society where equality of opportunity for participation in political and social practices can be assured need not be a bad thing (Silvers, 2008). Therefore, enhancement should not be rejected just on the grounds that it will diversify society. Nonetheless, we cannot generalise that more variety or diversity is intrinsically good. Diversity might be good, but it opens the door for a lot of things that not everyone approves of as good. For instance, people had great expectations about the diversity and freedom that online virtual worlds would bring, yet these places have turned out to be places where things that society does not generally support with open arms (such as pornography and betting) take place. The ways in which humans would alter themselves when more human enhancement interventions become available, and whether the alterations would truly add value or not to our lives and to a more diverse society, are still a matter for more research. In the meantime, it seems reasonable to question the power discourses that might underlie the diversity suggested by the transhumanist paradigm, a diversity that is not 'nature-given' or 'circumstance-given' or 'cultural-chosen', but 'individually chosen' and a 'technologically given' one.

Different views on the posthuman

While the particular view of the posthuman held by this paradigm might be problematic, the idea of the posthuman, broadly speaking, can still be worth exploring. As mentioned above, the question is not whether we will become posthumans, for posthumanity has been with us for a long time. Rather, the question should be, what kind of posthumans do we want to become? Posthumanity is, then, not so much about the future; it is also about the present—about how we live, the things we investigate, the questions we ask and the assumptions underlying them (Pepperell, 2004). In the previous section it was argued that the transhumanist view of the posthuman is not without problems. However, that is not to say that other forms of the posthuman are not worth pursuing (Cabrera, 2009b; Hayles, 1999).[43] For example, the view on the posthuman fostered by cultural posthumanism can help us to reach a wider acceptance of human differences and more tolerance, as well as better understanding the maps of power and identity that under-lie different bodies (Haraway, 1991; Hayles, 1999; Schneider, 2010). The posthuman need not entail the end of humanity as we know it, nor be the next logical evolutionary step. The posthuman could instead (1) indicate the end of a certain 'conception of the human' (Hayles, 1999), (2) be the realisation that the human only comes into existence by contact with others—being humans or nonhumans, which is also a view that is at the core of the concept of the individual as relational, (3) offer us interesting perspectives on our conception of the human condition—either by our anxiety or our enthusiasm about technologi-cal change, (4) help us develop a new sociology and a renewed form of ethics and (5) be an opportunity to get out of some of the old boxes and open up new ways of thinking, and understand the meaning of being human and leading the good life.

The human prejudice

'Humanity is in peril', not from the menace of our technological powers or from different sorts of existential risks, 'but from a conceptual threat' (Fernández-Armesto, 2005, p. 1) that comes from our particular under-standing of being human. As Judith Halberstam and Ira Livingston have argued, 'You're not human until you're posthuman. You were never human' (Halberstam & Livingston, 1995, p. 8).

It is not an exaggeration to say that most individuals in Western soci-eties take for granted a series of fundamental, unproven assumptions about human uniqueness compared to other living beings (Fernández-Armesto, 2005; Harris, 2010; Savulescu, 2009; Singer, 1975). Different arguments have been offered to support this view; for instance, that humans are made in the image of a divine supreme being, have a capac-ity to reason, engage in complex social relationships, display empathy and sympathy, and have a capacity to act from normative (including moral) reasons. However, none of these arguments have proved to be a compelling normative reason for privileging the human species. Moreover, as Harris has noticed, 'there is a long established and deeply ingrained habit, of identifying properties or qualities that are contingently possessed by human beings as necessarily possessed by our kind and, moreover, necessarily not possessed by other kinds' (Harris, 2010, p. 10).

This view is a bias which philosopher Bernard Williams called the human prejudice, and it has been referred to by others as the arrogance of humanism (Fernández-Armesto, 2005) and anthropocentrism. The human prejudice has promoted a 'separation of the human and the animal' (Campbell, O'Driscoll, & Saren, 2010, p. 88), making a difference to the ways in which we treat (and think we should treat) those that we regard as non-human beings. Some people have argued that such a position involves a form of speciesism—a view in which only human beings are worthy of moral consideration—which is as morally wrong as racism or sexism (Savulescu, 2009; Singer, 1975).

The transhumanist paradigm avoids such prejudice, as it does not start with the question whether this individual is a normal human being or is this a unique human being; not even with the question whether this creature is a human being at all. Rather, it starts with a question about the minimum set of traits that defines the things we consider worthy of moral consideration. For this paradigm, the answer to the latter question involves the concept of personhood. Surely an approach based on personhood has its own problems, such as the common intuition that we should care for humans even when they do not meet the personhood criteria (such as severely impaired humans) because we share with them something that we do not share with other beings. Another caveat with a personhood approach is that once we start the process of extending the threshold to include other non-humans in our circle of moral concern, it is not obvious where to stop. Once other beings are included on the grounds that their features resemble features that we consider worthy of moral consideration in humans, why not extend moral consideration to other beings that resemble the first beings to which we granted moral consideration?

Regardless of whether or not we embrace the personhood approach implied by transhumanism, its rejection of human prejudice is something we can learn from, and start to question whether we value human beings for being humans or whether we value the characteristics that make them persons. Moreover, rejecting the human prejudice can help us to recognise and rethink the way we treat each other, different species and the environment.

Conclusion

This chapter started by discussing the ideas underlying and shaping this human enhancement paradigm, namely transhumanism and posthumanism. These two movements shape the ways in which this paradigm envisions the use of emergent technologies, bringing to the fore different paths to improve ourselves via technological fixes. The particular features, views and values that the transhumanist paradigm holds were also discussed. The chapter then examined the set of arguments used by this paradigm to support enhancement, as well as the several controversies that the paradigm confronts. The chapter ended by discussing some of the valuable ideas that the paradigm brings forward, such as the idea of the posthuman, if explored from a cultural posthumanism perspective. This paradigm, as well as the biomedical one, falls short of acknowledging that human enhancement need not be only a technological *individualistic* endeavour, but that it can also be a *social* endeavour. Furthermore, it might be the case that we still do not have a better framework for evaluating enhancement ideas that do not necessarily share all the commitments held by the extreme standpoints—bioconservatism or transhumanism. It is in this spirit that the next chapter, inspired by a more inclusive and relational view, explores a suggested third human paradigm.

4
The Social Paradigm

A social approach

Previous chapters have discussed two common paradigms within the human enhancement discourse, the biomedical and the transhumanist. This chapter introduces a third paradigm, one that has been motivated by the acknowledgement that any comprehensive debate on human enhancement should begin by recognising the rich and complex relations that shape who we are as well as the social factors that currently contribute to our understanding of what it means to be healthy, the goals of medicine and what it means to be human. This paradigm also recognises that the scope of human enhancement goes beyond the relationship of humans with medicine and its purposes, and beyond the relationship of humans with technology.

Something new on the horizon?

There are already different views of and approaches to human enhancement, thus it can be questioned whether or not there is anything original or novel about a social perspective on the subject. However, it is precisely this combination of a social perspective and enhancement that has not been fully explored or discussed. The different human enhancement approaches suggested so far have started from the premise of the individual as an abstract and isolated agent (the liberal view) without proper recognition of the real nature of individuals, namely that they are relational and dependent on others' beings (the relational view). These human enhancement approaches have also been dominated by high-technological individualistic interventions focused on changing an individual's bodily and mental features, and mostly focused on the interests, desires and values of a small group of privileged individuals (mostly Western white men with certain economic advantages).

Considering all this, the suggested third paradigm brings new aspects that have been undermined in other human enhancement paradigms. The approach suggested by this paradigm is new insofar as it offers a view on human enhancement in which the relationships shaping who we are as well as our individual lives are not only the starting point for reaching human enhancement, but determine the conditions needed to reach it. By highlighting a broader and different set of interests, desires and values, this paradigm might be doing some remedial work—adding those who have been left out of the enhancement discourse—but more importantly it urges us to rethink the assumptions upon which the current human enhancement discourse is based. Furthermore it urges us to consider the possibility that far from being a source of enhancement, the principles, values and criteria promoted by the biomedical and transhumanist paradigms of human enhancement actually reinforce patterns of domination and subordination that lessen human well-being. This paradigm offers us an opportunity to move forward and bring insight to the human enhancement debate, as it seeks to respond to the lack of attention regarding the relational nature of individuals under the current dominant human enhancement paradigms. A relational view of individuals could help us to balance individualistic preferences with social needs. Thus, social enhancement brings to the fore a more inclusive, equitable, sustainable way in which human enhancement can be understood and pursued.

Having a particular paradigm focused on the social is not without purpose. Paradigms can be understood as powerful tools that can change the way we understand the world and the problems we encounter in it, tools that structure and shape our actions (Clark, 1998; Kuhn, 2012; Wittgenstein, 1953). The result of living in a society in which enhancement is always associated with the posthuman as the result of radical technological interventions, with injustice, with individuals having super powers,[44] compared to one where human enhancement promotes social transcendence rather than individualistic transcendence,[45] and focuses more on our relatedness to others and the world rather than on us becoming more like machines, will be substantively distinctive. Different discourses can serve not only as an instrument for achieving different goals (such as social enhancement), but as Wittgenstein expressed it (1953), to have a framework to counterbalance those discourses that are already out there. Moreover, words have meaning because they create a 'common vantage point from which we survey the world together' (Taylor, 1985, p. 259). Thus social enhancement can be seen as a novel way in which to explore and reconceptualise human enhancement, a discourse that reminds us that there are alternative views to the current understandings of human enhancement. At the same time, social enhancement can help us to interrogate the ways in which human enhancement has been understood, the values it has promoted and the relationship of forces it has involved.

In order to better grasp the suggestions made by this human enhancement paradigm, the next section will present and outline two key concepts underlying social enhancement, namely social justice and the social determinants of health.[46]

Social justice

Justice is a complex and contested concept. There are multiple domains of justice and distinct views of what constitutes justice. Within each theory of justice there is substantial diversity with respect to what are the appropriate objects of justice, who are appropriate subjects of justice, and the relationship between justice and other ethical concepts. In the case of social justice, it depends on the criteria used: need, effort, equality of something or a combination of these. When talking about justice, and especially about distributive justice, it is common to think about the theory developed by John Rawls. In his seminal work *A Theory of Justice* (1971), Rawls argued that the main idea of distributive justice is that members of society, under a veil of ignorance and in a social contract scheme, would choose principles for the distribution of the primary social goods. Rawls uses the concept of a veil of ignorance to refer to the idea that members of society 'are not allowed to know the social positions or the particular comprehensive doctrines of the persons they represent' (Rawls, 2001, p. 15); thus they would not know their race, ethnic group or their various native endowments, such as intelligence or height (Rawls, 1971). By a social contract scheme Rawls is stressing the most fundamental idea in his conception of justice, the idea that society is a fair system of social cooperation over time. Finally, by primary social goods, he is referring to 'things citizens need as free and equal persons living a complete life' (Rawls, 2001, p. 58), including rights, liberties and opportunities, income and wealth and the social base of self-respect. Other important aspects of Rawls's theory of justice are the principle of liberty and the 'difference principle'. The former is concerned with each person having 'an equal right to the most extensive basic liberty compatible with a similar liberty for others' (Rawls, 1971, p. 303), while the latter allows for inequalities, as long as they work to make those who are worst off as better off as possible, compared to alternative measures and without undercutting equality of opportunity.

Rawls takes primary social goods as the embodiment of advantage, in which the aim of justice is to compensate for the outcome of the natural lottery. Thus, Rawls's approach implies that people with expensive or offensive preferences have no claims of justice on us. Daniels (1985, 2008) took Rawls's line of argumentation to defend his view on our duty to attain species 'normal functioning' as the path for equal opportunity. It is questionable, however, that normal functioning is necessary or sufficient to give individuals an equal range of opportunity, and even less that we have any obligation to promote normal functioning. Even if we agree with the idea that there is such a thing as a normal range of opportunity (later Daniels changed this from the normal range to the effective range), that does not mean that individuals need to be within 'normal' parameters to achieve the same ends. Consider, for example, the case of double amputee individuals, who are generally considered to be outside 'normal' parameters regarding bodily appearance. These individuals have shown us that in order to be able to move from point A to point B it is not necessary to have two human legs, but rather robotic legs, an extra leg (as people who are assisted by a walking stick) or even no legs at all (such as someone in a wheelchair) can also get the task done. Even though Daniels has acknowledged that the normal opportunity range is social relative, dependent on various facts about the social organisation and society's level of technological development, his approach seems to undermine that the same happens when we talk about the normal functioning of individuals.

According to Daniels (2008), Rawls's opportunity range is also similar to the neo-Aristotelian capability approach suggested by Amartya Sen and Martha Nussbaum. According to this, a capability is an exercisable or accessible opportunity or option needed to function as a cooperative member of society (Nussbaum & Sen, 1993; Sen, 1979). Thus, the approach has been regarded as fitting with general and widely shared intuitions about human well-being. Unlike Aristotle's approach, this one is not intended to be comprehensive but partial, a moral conception selected for political purposes.

Taking into consideration that Sen and Nussbaum's capability approach is focused on a sufficient set of capabilities, and not on primary social goods, it is not clear how this approach is comparable to Rawls's approach. Furthermore, it is not only that they are focused on different things, but also that the capability approach is trying to address some of the weaknesses found in Rawls's approach. A focus on primary goods, as Sen has argued, suffers from being concerned with goods, rather than with what these goods do to human beings (Sen, 1979).

Furthermore, Rawls's approach neglects the fact that taking advantage of primary social goods depends on a 'relationship between persons and goods' (Sen, 1979, p. 216). Persons have different biological realities and needs varying with factors such as age, climatic conditions, work conditions, body size and even beliefs. Considering this, we can see why an index of primary goods not only fails to capture certain hard cases (such as cases involving individuals with certain body or cognitive impairments), but also fails to notice very widespread and real differences (Sen, 1979). It makes more sense, then, to focus on whether individuals have the appropriate set of capabilities to do or be what they choose (Nussbaum & Sen, 1993; Sen, 1979). Capabilities are so fundamental to what we can do or can be that they seem to be a suitable starting point to discuss social justice concerns regarding human enhancement. From this perspective social justice is focused on the idea of capabilities as promoters of equality of opportunity; opportunity not only to contribute as a member of society but also to select, revise and count with a social and political environment to achieve our goals and ends. Here it is important to bear in mind that in pluralistic societies, where the definition of what is a good human life is likely to vary, we cannot commit ourselves 'to pursue equality of capabilities, but only assuring that individual's capability sets are not distinctly worse than those of others' (Daniels, 2008, p. 68; c.f. Nussbaum & Sen, 1993).

Underlying the capability approach is the idea that most reasonable people would agree that all living beings worthy of moral concern share a set of basic capabilities ($C_1 \ldots C_n$) that impact their quality of life. Some capabilities in the set might be idealised and only valid for the majority (but by no means all humans), but if you do not have or exercise the essential ones (such as a minimal level of consciousness or certain brain functions) then your status as a being worthy of moral concern is on the line or you are simply no longer alive. In the case of human beings, even though people might have very different needs varying with health, longevity, climatic conditions, location and work, there are still some biological needs, at least in principle, that are similar among those we recognised as humans (Sen, 1979). If this is true, there must be ways in which to secure that beings worthy of moral concern enjoy those basic capabilities. Failure to meet these basic capabilities is a problem of justice (Nussbaum, 2001).

The capability approach focuses on promoting capabilities and not actual functioning (abilities); in that way the individual can choose to pursue or not pursue the relevant functioning (Nussbaum, 2001; Robeyns, 2006). An emphasis on capabilities rather than actual functioning allows for different conceptions of the good life to be taken into account; it allows for individuals to take responsibility for their choices, and it allows us to consider the value of a social level of capabilities. Accordingly, it is not the attainment of 'normal' features that this approach is interested in; rather it argues for equity in the space of opportunity, as this would allow individuals to perform as active members of the community.

The approach is not without its own concerns; for instance, it is debated which capabilities should be considered as basic. While Sen has explicitly refrained from defending a well-defined list of capabilities (Robeyns, 2006), Nussbaum has come up with a list of 'central human capabilities' (2001). She puts life as her first central human capability. She argues that this central capability implies '[b]eing able to live to the end of a human life of *normal* length; not dying prematurely, or before one's life is so reduced as to be not worth living [emphasis added]'.[47] It is not clear what she means by normal: if by it she means similar to the standard set by Western countries, which would be close to the biomedical use of the term, then this would be reinforcing an arbitrary account of normalcy. That is one of the reasons why continuous social deliberation about the capabilities we want to have included within the basic set of capabilities is needed. Ingrid Robeyns (2006) has explored some options for arriving at a consensus of the capabilities to be included, while at the same time emphasising the need to reach agreement regarding the order of aggregation of different capabilities. Despite these concerns, the capability approach is a good starting point in our discussion of justice, not only because it acknowledges pluralism and cultural difference as well as the complex and rich relationships between human striving and its social and material context, but also because it promotes capabilities based on real need rather than mere desire.

Social determinants of health

People's health and well-being are shaped by social and environmental factors and by social and economic policies (Beaglehole, Bonita, Horton, Adams, & McKee, 2004). Factors such as the state of our environment, our relationships with others, where we live, our genetic make-up, education level and income all have considerable impacts on our health and well-being, whereas factors such as access and use of medicines or health care services often have less impact than is commonly attributed to them. This knowledge is not new; for instance, most of the modern reduction in mortality from infectious diseases took place prior to the development of effective biomedical therapies, by improvements in sanitation, living conditions and changes in food supply (Irwin & Scali, 2010). To quote a recent WHO report:

> Water-borne diseases are not *caused* by a lack of antibiotics but by *dirty water*, and by the *political, social, and economic forces* that fail to make clean water available to all; heart disease is caused not by a lack of coronary care units but by lives people lead, which are *shaped by the environments in which they live*; obesity is not caused by moral failure on the part of individuals but by the excess availability of high-fat and high-sugar foods.
>
> (CSDH, 2008, p. 35 [emphasis added])

The social, physical and economic environments in which people are born, grow, live, work and age, are referred to as the social determinants of health (SDH). These reflect people's different social position in terms of power, income, resources, status and services (Akerman, 2009; Blas & Kurup, 2010; Irwin & Scali, 2010). The SDH are constituted by structural determinants and the conditions of daily living. The former are defined by social stratification and its sustenance mechanisms in society, including the nature and degree of social stratification; biases, norms and values within society; economic and political participation; global and national economic and social policy (such as social exclusion, income and wealth distribution, education, non-regulated markets and advertising); and process of governance (civil rights, employment conditions, public spending priorities; health care system; and macroeconomic conditions). On the other hand, the conditions of daily living are related to specific factors and circumstances, such as community settings and infrastructure, psychosocial circumstances, working conditions, and biological and behavioural factors.

While there are different approaches towards the SDH, this book follows the ecosocial approach, which brings together the psychological and social production of health approaches and explains inequities in health across socio-economic positions as the cause and the result of dynamic, ecological and historical relationships (Akerman, 2009). The value of this SDH approach is that it offers an integrated view of the changing patterns of health within communities by grasping 'how social and physical environments interact with biology and how individuals "embody" aspects of the context in which they live and work' (Kelly, Morgan, Bonnefoy, Butt, & Bergman, 2007, p. 11).

The SDH discourse, as well as the capability approach, promotes a view of the individual that better grasps the richness and complexity of the relationships between our well-being and our environments. This is exactly the perspective that the social enhancement paradigm has at its core, namely a relational view of the nature of individuals.

A third paradigm: Social enhancement

A liberal view of the individual, in which individuals are seen as abstract, independent, rational and self-interested agents, each with his or her own private interests that must be respected and accommodated as far as possible, has permeated contemporary Western societies (Held, 2001, 2006), including the biomedical and transhumanist paradigms of enhancement. While this view can be useful for certain purposes, such as certain market policies, 'we should not lose sight of [its] restricted and artificial aspects' (Held, 2006, p. 28). Individuals are neither discrete nor circumscribed beings. Sartre once said that 'man is defined by its relation to the world and his relation to himself' (2012). Who we are and who we will become is shaped and given meaning to a great extent by the rich and complex relations we have with other individuals, our environments and communities.

This view about the nature of individuals is lacking in both the biomedical and the transhumanist paradigms of enhancement. The social enhancement paradigm attends to the relational nature of individuals, building awareness and calling for mutual responsiveness. It aims to foster a new way of thinking that begins with the premise that we live not in separation but in relationship (Gilligan, 1993). It emphasises a view in which society is not the sum of isolated entities who do not contribute to each other's well-being, but rather it sees society as the result of complex and rich relations between individuals in their community. Finally, it acknowledges that sustainable and meaningful enhancement is something that cannot be attained by merely focusing on individualistic interventions.

Social enhancement

While the previous two paradigms were focused on individualistic types of intervention, albeit with different scopes of individual change, the social paradigm holds the view that meaningful and sustainable enhancement can only be achieved by acknowledging the rich and complex relationships shaping who we are and our well-being.

This is not to say that individual interventions are not beneficial, but rather that they are not the only available option (nor the most appropriate one in most cases) to improve the well-being of individuals.

Under this paradigm, enhancement options are neither confined to medical solutions or technological gadgets, nor to direct interventions in the human body. Just as wealth alone does not have to determine the health of a nation's population (CSDH, 2008), high-technological enhancement interventions do not have to determine improvements in the human condition. Take the example of Cuba where, without recurring to high-technological interventions, the country has maintained a reduced child mortality rate and a reduced number of women dying from childbirth (CSDH, 2008). Or consider the case of children who develop in more supportive environments and with the necessary levels of nutrients who have a lesser risk of cognitive impairments than children in less supportive environments. The interventions to achieve these great social and community benefits are not attained through expensive, highly sophisticated interventions; rather they are attained by inexpensive, fairly safe, low-technology interventions that have taken into account the rich and complex relationships shaping our well-being. Thus, enhancing the social conditions in which humans develop, live and work might be as important, if not more so, than directly changing the biological features of individuals.

Considering all this, we can now move on to what social enhancement is.

Social enhancement: any intervention that augments or improves an individual's *capabilities set* with the aim to *enable* and *empower* them as *active members of society*, without directly changing the biological reality of individuals.

Social enhancement is therefore not about directly changing individuals and their onboard capacities, but rather about changing the environments that disable us from performing at our optimal levels, both physical and mental levels. Interventions aimed at changing our environments would include, for instance, interventions in the infrastructure, food and textile sectors. In addition social enhancement is focused on the enablement and empowerment of individuals as members of rich and complex relationship networks, mostly through collective action, cooperation and better understanding of our relational nature.

This paradigm takes insight from the SDH discourse in which 'social' is generally used as an umbrella term to include the political, environment, economic, cultural and psychological relationships influencing, in this case, enhancement outcomes and aims; thus social enhancement. By referring to this paradigm as social a different focus is emphasised than in the biomedical and transhumanist paradigms, namely the social and communal level rather than the individual level; but more importantly it emphasises a completely different assumption about the nature of individuals. However, 'social' does not refer to an enhancement of social capital, which is commonly understood as the connections within and between social networks; yet improvements in social capital can be a consequence of social enhancement. In addition, the fact that a given enhancement intervention is widely available is not enough for it to be counted as social enhancement. Even if a given intervention were focused at the societal level, as long as its main mechanism of action is changing individuals, it would not count as social enhancement. Finally, social enhancement is closer to Popper's idea of piecemeal engineering than utopian social engineering (1961), as social enhancement seeks to better design our environments to empower individuals rather than trying to remake society from the ground up. Moreover social enhancement interventions can be implemented democratically; that is, openly debated.

The paradigm

One important feature of this enhancement paradigm is that it holds a different assumption about the nature of individuals. Contrary to the abstract and separate individual assumption from which the other two paradigms start, this paradigm takes individuals to be relational and dependent (Held, 2006). Social enhancement interventions do not promote individualistic values and desires (such as the goals of getting better than well or reaching a posthuman stage); rather they are focused on promoting social values (such as cooperation, caring and empathy).

Another important feature of this paradigm is that human enhancement interventions are neither seen as intrinsically morally wrong (biomedical paradigm) nor as intrinsically morally good (transhumanist paradigm); rather their moral worth is context-based and needs to be assessed case by case. While enhancement under this paradigm is not based on a definition of health or disease, it is still concerned with attaining health and well-being, but compared to the other two paradigms the suggested paths of action are quite different. In this regard, this paradigm has similar paths of action to those of modern public health frameworks, such as the promotion of well-being through community effort and action, and acknowledgement of the role of the environment and social machinery. The following are some factors that influence the specific paths of action on which this paradigm is focused:

- The way power and resources are distributed. That is why social enhancement promotes interventions that are not focused on changing directly the biological reality of individuals, but rather on social and community-based interventions.
- The deficiency in settings and availability of services affecting communities through time. Therefore, interventions need to be carefully chosen *according to context* in order to ensure that their benefits can be *sustained through time* (are both available and accessible) and are *sustainable* for the community (including appropriateness and affordability).
- Values, culture and beliefs that influence well-being. Thus social enhancement offers an alternative set of values, such as cooperation and community-oriented values (such as empathy and solidarity), in which human enhancement can be framed.

Under this paradigm, enhancement interventions are seen as *improvement for*, in contrast with the *improvement of* approach of the other two paradigms. Thus, it is not focused in interfering directly with our body's architecture or biology, but rather it promotes interventions that aim at exploiting the causal pathways that can change how information is accessed and processed (Nordmann, 2004). This paradigm does not start from the assumption that the only way to enhance individuals is by direct manipulation of their biological reality; rather it starts from the assumption that the different relations we have with others and our environments shape our well-being. Thus by shaping our environments in the right way we can enhance our well-being. Social enhancement, together with other modern disciplines such as epigenetics and environmental health, acknowledges the role the environment plays in our general well-being. There is already ample evidence for the impact that one's environment has on health and well-being over the course of one's life (Blas & Kurup, 2010; CSDH, 2008; Prüss-Üstün & Corvalán, 2006), as well as evidence showing that brain development is highly sensitive to external influences in early childhood, with lifelong consequences (Grandjean, 2013). Even though certain changes in the environment might not seem to bring any benefit to certain groups of people, even minor changes could make a major difference for others, such as young individuals in development or people with different biological realities from those that are accepted as the norm for humans.

There are two main reasons for this paradigm's position regarding direct manipulation of individuals: first, the fact that community members have different abilities, biological realities and preferences; second, the importance that different relations have in defining who we are as individuals and which contribute to our improvement as social and relational individuals. In connection to the latter, social enhancement acknowledges individuals' differences in internal capacities and social situations. That is to say, different members of the community are not equally able to convert and use the same interventions as forms of enhancement, and contributing enhancing factors work through complex interactions and relations. For example, for any trait, or function, there is likely to be a discrete range of optimal levels for the given set of conditions, and too much or too little may detract from health or well-being. This paradigm takes on board the idea that as part of our relational nature we are embodied, embedded and extended nature, and argues that this provides good reasons for us to move the focus of attention, in relation to enhancement, from directly changing our capacities to changing our environments.

As part of addressing key aspects of our environments, social enhancement is focused on creating environments—physical, political and social—that can be exploited by different members in a way that empower individuals. Shaping our environments can have more direct and effective impact to our well-being than higher-technological enhancement interventions (Blas & Kurup, 2010; CSDH, 2008; Levy, 2012). By changing our environments in positive ways, we are improving our well-being and the range of opportunities open to us.

The idea of influencing the environment to improve development and function is not a new one (Santiago Ramon y Cajal and Andy Clark have written extensively on this). Clark has said, for instance, that 'we may often solve problems by "piggy-backing" on reliable environmental properties' (Clark, 1998, p. 45). Some even argue that the option of changing our environments is something that we have already been doing. While this might be true, we can also say that many of these changes have created ill environments (such as information-loaded environments) that are detrimental to our well-being.

There is evidence pointing out that approximately 24% of the global disease burden and 23% of all deaths can be attributed to environmental factors, as well as the fact that urbanisation when poorly planned reshapes population health problems (increasing the amount of obesity, non-communicable diseases, accidental injuries, greater impact from ecological disasters) (Blas & Kurup, 2010; Prüss-Üstün & Corvalán,

2006). Thus, one of the aims of social enhancement is to enable environments that are not harmful and more importantly that allow individuals to flourish and improve their well-being. Bringing the right kind of environments is not only a feasible option (Marshall, 2010; Prüss-Üstün & Corvalán, 2006), particularly considering that we have the technology and knowledge to do so, but is also a very desirable option, as it is mostly based on interventions that are overall more equitable, socially oriented, relatively simple, fairly safe, non-invasive and inexpensive.[48] Adequate environments are important because they enable the conditions needed for people to live the lives they value (CSDH, 2008) and to empower communities.

The last key feature of this paradigm has to do with an idea underlying Clark and Chalmers' extended mind thesis, namely the *parity principle*. This principle suggests that as we confront a given task the part of the world that functions as a process that could go on as well in the head (cognitive process) becomes part of the cognitive process (Clark & Chalmers, 1998). Their thesis is that cognition is beyond the limits of skin and skull. For our purpose here we do not need to accept Clark and Chalmers' extended mind thesis in full, it is sufficient to agree with their views about extended cognition and active externalism (the active role of the environment in driving cognitive processes). Both of these views are compatible with the view that truly mental states—experiences, beliefs, desires, emotions and so on—are *emergent* brain processes.[49] Building on this feature of the social paradigm, if we take the Tetris example mentioned in Chapter 3, the preferred option for rotating the object on the screen would be the press-button. This would be the preferred option under this paradigm not only because it allows us to do quicker calculations than just using our heads, but also because it is a not-expensive and fairly safe environmental tool that presumably can be used by anyone who sits in front of that computer rather than just by a specific individual using the neuro-implant. At the same time, the press-button option can help mitigate certain cognitive and perceptual problems that individuals might face when performing mental actions (Kirsh & Maglio, 1994).

To sum up, social enhancement's view of the nature of individuals as relational offers us a promising alternative view to the enhancement views based on the dominant liberal approach, moving from an individualistic pursuit to a collective one. A view of human enhancement as suggested by the social paradigm would therefore be committed to invest in sustainable policies (integrating public, private, nongovernmental and international organisations and civil society), in infrastructure and actions to address the determinants of health and community development, and in protection from harms and the enablement of equal opportunities for health and well-being for all people. This changes the ways in which ethical problems are interpreted, what many think the recommended approaches to ethical issues ought to be, our actions and intentions, as well as the ways in which we develop and use our technologies.

The role of emergent technologies in the social paradigm

While the uses of technology in the biomedical and transhumanist paradigm tend to eclipse the social dimension of health and well-being, the social paradigm embraces this social dimension. Moreover, under this enhancement paradigm the role of technology is not so much a tool capable of distorting human nature, but rather one that enables individuals and society to find opportunities to improve the human condition, fulfilling the licit desire of a better life in a respectful and ethical way.

The main role of emergent technologies under this paradigm is therefore focused on positively changing the social and environmental conditions in which we are born, live and work, thus helping to address the SDH. Taking this into consideration, one feasible and reasonable option that could yield substantial results is environmental interventions, in particular acting on *modifiable* environmental factors that are realistically amenable to change by using current technologies.[50] Here, emergent technologies will be used with a focus on creating fair, sustainable, accessible environments that enable us as individuals and members of communities to develop and make optimal use of our given capabilities. These environmental factors include physical, chemical, biological and even cultural and political features of the environment that directly affect well-being and behaviours (Prüss-Üstün & Corvalán, 2006).

An example of a social enhancement intervention is sOccket, a soccer ball which charges a battery every time it bounces, providing light to children living in places where there is no light infrastructure while helping them engage and exercise as part of a community.[51] This is an example of social enhancement intervention as it is not enhancement *of* the child, but rather is enhancement *for* the child, enabling an environment where children can exploit their own capabilities, not only on the playing field but at home, where they can use the light provided by the ball to study, read and so on. It is also an example of social enhancement as it does not need many resources to be used or implemented, does not involve high-sophisticated and radical technological interventions, and is sustainable, as it can be used to address global energy issues. Moreover, it is fairly safe, non-invasive, promotes healthier behaviour (playing and exercising) and empowers individuals.

Other examples of social enhancement in which emergent technologies can be used are the following:

Biological and behavioural factors

Another simple way of social enhancement is by interventions in food sources to ensure the necessary level of micronutrients in people's diets (such as iron, iodine and zinc). Micronutrients are important because, according to scientific studies, not only do they improve mental capital but they also avoid cognitive impairments and mental disorders (Blas & Kurup, 2010; Gómez-Pinilla, 2008; Irwin & Scali, 2010). Micronutrients deficiency is one of the most prevalent nutrient deficiencies in the world, affecting an estimated two billion people, and causing almost a million deaths a year (WHO, 2002). A recent example of how emergent technologies can help in this area is illustrated by research from ETH Zurich and the University of Zurich, in which a series of nanostructured iron and zinc complexes were shown to be well absorbed by rats without modifying the colour and taste of some of their common foods (Miller, 2010). If these nanocomplexes were to pass bioavailability and efficacy in humans they could be a feasible and a cost-efficient method to ensure people's necessary micronutrient levels.

Chemical factors

Environmental interventions focused on chemical factors could include smart sensors that can detect, for instance, dangerous chemicals in the air, water or food, and changes in body chemicals.[52] Smart sensors— whether located in food, textiles or the environment—could provide a cost-efficient option to help individuals and other members of the community to better respond to changes in their bodies, environment and even social context. This relates to Kris Pister's idea of 'smart dust', tiny sensors—small as particles and acting like electronic nerve endings for the planet—that would make observations, monitor and relay mountains of real-time data about people, cities and the natural environment (Ilyas & Mahgoub, 2006).

Specific examples in which emergent technologies are making and could make a contribution in this area include:

(1) Sensors for smartphone applications that can track airborne toxins or pollution in real time, alerting the individual and emergency responders (Baker, 2011).

(2) Smart textiles offering protection from hazardous chemicals, heat, extreme cold and radiation, and with specialised interconnectivity to online applications (Joshi & Bhattacharyya, 2011).
(3) Smart paints that change their colour when certain chemicals are detected in air or soil.

Smart sensors go beyond measuring chemical parameters to areas such as brain waves, heart rate, temperature, stress level monitoring and context awareness recognition.

Community settings and infrastructure

To a certain extent, a large number of people already live in enriched environments, with access to sanitation, electricity and green areas. However, there are many people who live and work in ill environments. Enriched environments have been shown to enhance cognitive function in various learning tasks (During, Young, Lawlor, Leone, & Dragunow, 1999; Mora, Segovia, & del Arco, 2007; Rampon et al., 2000; van Praag, Kempermann, & Gage, 2000). There is also evidence that environments promoting more aerobic exercise bring benefits to our cognitive performance and well-being (Scholey, Moss, & Wesnes, 1998). Thus, creating healthier environments has the potential to enhance or at least protect our well-being and mental capital.

Other examples of enriched and smart environments that are already available are the World Wide Web, virtual reality interfaces and some basic augmented reality interfaces. In the case of the Internet, it is becoming a primary resource for keeping track of information, keeping people and the world connected, and bringing literacy to individuals who cannot attend traditional schools or live in remote areas. Among the advantages of the technologies and applications enabled by the Internet are their accessibility (they can be found and used in low- and high-income countries and in urban and rural areas), affordability and safety (particularly compared with other enhancement applications). Thus, web-based applications can be considered as one of the most realistic and already widely available means for social enhancement.

In the case of virtual reality, the metaphorical spaces that arise through interaction with machines (generally computers), in which people navigate, in some cases using special hardware—such as helmets, goggles and data globes—but in some cases only by using words, can be an opportunity for 'experiential and active learning in challenging but safe and ecologically-valid environments' (Weiss, Rand, Katz, & Kizony, 2004). Here we can also mention computer-based training programmes as well as video games developed specifically for improving domains such as attention, executive control, processing speed and visuospatial skills (Jak, Seelye, & Jurick, 2012).

Finally in connection with augmented reality, in which virtual and physical surroundings are blended, we can distinguish between individual-based augmented reality and social-based augmented reality. The former, which layers virtual imagery and information over real-world environments for an individual, is already available in most smartphones and other industrial and web applications (such as Google maps) by integrating information from inexpensive sensors, global positioning and recognition systems. The latter is related to what is known as 'ubiquitous computing' or 'ambient intelligence', in which a whole infrastructure (for instance a building or a park) would integrate information from different kind of sensors (such as movement, temperature, chemicals, recognition) and use this information to respond to specific situations, bringing augmented reality for anyone in that environment. Within a community, for instance, this type of social augmented reality system could alert the members of the community about certain events, such as vegetables ripening, failures in buildings or toxic substance spills. Environments equipped with monitoring and feedback technologies could allow us to monitor and reason about our behaviours, goals and intentions, thereby facilitating access to information, communication channels and even the control of artefacts through speech recognition, tangible interfaces and display technologies ubiquitously embedded in our environments (Olivier, Wherton, & Monk, 2009).

The examples above show the commitment social enhancement has to interventions that are responsive,[53] diverse and with a long-term impact on well-being for the community and individuals, acknowledging our relational nature with other individuals and the environments in which we live and work. They also show alternative ways in which individuals and communities can be enhanced significantly without changing directly their biological realities.

To sum up, social enhancement interventions under the social paradigm are not about directly changing someone's biological reality, but instead they are about changing our environments. The fact that these interventions do not change directly our biological reality does not mean that changes would not occur. Research has shown that environmental interventions not only impact behaviour and social outcomes, but also bring significant changes at the brain connectivity, cellular and molecular levels. Social enhancement interventions are aimed at bringing improvements not from an individualistic perspective but from a relational one. Here the benefits of social enhancement interventions have been highlighted, but that is not to reject the possibility that social enhancement, just like biomedical and transhumanist enhancement interventions, could bring forward unforeseen and undesired effects.

The ethics of social human enhancement

The arguments

In general the arguments supported by the social enhancement paradigm tend to take a moderate approach between the views held by the biomedical and the transhumanist paradigms. But more important is the fact that a different set of values and a different perspective of the individual are behind the ethical stand of social enhancement. This section will point out the particular view this paradigm takes regarding the different arguments discussed for the other two enhancement paradigms.

Definition

Under the social paradigm of enhancement most therapies, if not all, are considered a form of enhancement, insofar as they serve at least the well-being of the individual; however, it acknowledges that not all enhancement interventions are therapeutic. In this regard this paradigm concurs with the transhumanist paradigm view in that often therapy and enhancement goals overlap, making it difficult, if not impossible, to draw a meaningful distinguishing line between them. While some scholars have argued that we should keep a distinction between therapy and enhancement for practical reasons, the distinction would still not be useful for mapping a boundary between morally obligatory and non-obligatory interventions (Daniels, 2000) or morally praiseworthy and non-praiseworthy interventions (Racine, 2010). To quote the President Council on Bioethics report: 'relying on the distinction between therapy and enhancement to do the work of moral judgment will not succeed' (PCB, 2003, p. 16). In addition to this, human capacities have already been enhanced, through various kinds of interventions and technologies, such as mobile phones, the Internet and computers; and as such it makes sense to expand the discussion around enhancement beyond the medical sector and the therapy-enhancement distinction.

Furthermore, given that this paradigm does not start from the assumption that individuals are abstract and independent beings, but rather relational beings, its definition of enhancement detaches from the common individualistic and direct bodily alteration view. Our well-being is shaped by our relationships with others and our environments; thus, enhancement interventions cannot keep undermining this fact. Finally, this paradigm recognises that what we associate with enhancement is closely connected to the things we value as individuals and as communities. Bostrom and Sandberg make the point when they argue that 'we might have no reason to value an enhancement of our sweat glands that increases their ability to produce perspiration in response to heat stimuli' (Bostrom & Sandberg, 2009a, p. 378). However, in certain environmental conditions such an intervention could indeed be considered to be a valuable intervention.

Naturalness

The social paradigm acknowledges that concepts such as normal and natural are changeable, value-laden, subjective and arbitrary. The term 'normal', for example, can refer to the individual or to a certain society, or it can be understood from a species perspective, a biological perspective or a cultural one. For instance, what does it mean to say that someone has a normal body, or a normal gene? Or consider the different conceptions of what normal entails if contrasting Western with non-Western, or abled with 'disabled' perceptions. Accordingly, this paradigm avoids entangling itself with these concepts, as these have failed to provide sound arguments around the moral permissibility of enhancement interventions. It also avoids the power issues that underlie who decides and how it is decided what being healthy, normal or natural is.

Under this paradigm it is also acknowledged that we are technological beings as much as we are biological beings; thus enhancing ourselves through technological means is not seen as artificial or necessarily incompatible with our grasp of the purpose and meaning of humankind and the universe. However, even if it is natural that humankind has pursued enhancement throughout history it does not follow that any kind of intervention promising to be a path to improve the human condition is acceptable or should be counted as an instance of human enhancement, nor that humans should regard themselves as the masters of nature. Thus under this paradigm the transhumanist view about humans taking control of their own evolution through radical technological interventions that directly affect their bodily and mental features is rejected.

Autonomy and freedom

The social paradigm agrees with the biomedical paradigm in that autonomy can be threatened by interventions that stop us from searching for a meaningful understanding of ourselves. It also accepts that autonomy and freedom can be threatened if they are restrained, coerced and manipulated by those in power in order to serve their own interests rather than the common good. Simone de Beauvoir captured this idea when she claimed that:

> It is other men who open the future to me, it is they who, setting up the world of tomorrow, define my future; but if, instead of allowing me to participate in this constructive movement, they oblige me to consume my transcendence in vain, if they keep me below the level which they have conquered and on the basis of which new conquests will be achieved, then they are cutting me off from the future, they are changing me into a thing.
>
> (de Beauvoir, 2011, p. 82)

The particular views on autonomy and freedom held by the social paradigm make them incompatible with the views held by the transhumanist paradigm. The transhumanist view places autonomy and freedom at the top of its pillars and as the basis for individuals to employ and seek enhancement interventions based on their own judgement of potential benefits and risks. From a social paradigm view such understanding of autonomy and freedom is ill-founded. It is ill-founded because it assumes that all people live in liberal democratic societies with access to similar levels of education and enjoyment of freedoms. Surely, that is not the case. Many of the freedoms and goods enjoyed in our society are not universal, and as such we cannot presuppose that the kind of enhancement interventions suggested by the transhumanist paradigm will benefit people living in less (or un-) democratic societies or with less access to education and wealth. Another reason for rejecting the transhumanist view is that even if we grant the view that autonomy and freedom are important, this need not mean that when in conflict with the freedom of others they should override concern for others' well-being. Freedom of choice should not be confused with freedom to do. Here we can mention John Stuart Mill's harm principle (Mill, 1863), which states that as long as our actions do not result in harm to non-consenting others or to the individual's self we are within our right to perform that action; otherwise we ought not to act. However, even if most people accept that the pursuit of well-being should be constrained by this principle, it must still be recognised that unless we take seriously the role our relations and context have in shaping who we are, the pursue of autonomy and freedom can be harmful to our communities and even ourselves.

It is from this last point that the real problem concerning human enhancement in relation to autonomy and freedom arises. Both the biomedical and the transhumanist paradigms' position on autonomy and freedom (albeit in their own particular perspective) presuppose liberal political thought. The problem with such a predisposition is not that it makes the debate on human enhancement political; rather the problem is the particular view in which autonomy and freedom are based, namely a liberal individual view.

Freedom and autonomy, from a relational view, have to do with lives lived in relationship, interdependence, context and connection; and right relationship is integral to the capability of both persons and natural and social systems to function and flourish. Relational autonomy embraces the fact that individuals are inherently social, political and economically situated beings; it encourages us to pay close attention to the type of forces that shape our decisions, rather than pretending that we can make decisions 'free' of outside influences (Baylis, Kenny, & Sherwin, 2008). Thus, the autonomy and freedom sought within the social paradigm have to do with 'a capacity to reshape and cultivate new relations, not to ever more closely resemble the unencumbered abstract rational self of liberal political and moral theories' (Held, 2006, p. 14).

Dignity

Within this paradigm the concept of dignity is not regarded as useless, but it is also not considered as something specially granted only to humans. This paradigm holds, instead, that dignity reflects a certain quality, a kind of excellence admitting of degrees and thus applicable to living beings within and without the human realm (Bostrom, 2007; Glenn & Dvorsky, 2010). Accordingly, within this paradigm dignity is not only granted to human beings, but also to non-human beings (such as other animals and nature in general). Moreover, both enhanced and unenhanced living beings can have dignity, albeit perhaps with different degrees or even kinds of it. However, even when scholars such as Bostrom have argued that we can extend dignity to entities (including technological entities), they have failed to establish the reasons for us granting dignity to mere technological devices (Jotterand, 2010). So for instance, while we could talk about a particular case of human or posthuman dignity—as reflecting his or her quality as a biological, social and technological individual—it would be awkward to say that my cochlear implant can be granted with dignity. Nonetheless, here we are entering murky territory, because if we ever managed to create artificial conscious life, it is not clear whether this would have the kind of quality we generally associate with dignity.

Finally, contrary to the biomedical paradigm, social enhancement does not consider technological enhancement interventions per se as threats to our dignity. Nor does it consider that enhancement interventions which do not reflect the *individual's autonomous* choice to enhance are undignifying as held by the transhumanist paradigm. Instead, this paradigm holds that interventions that neglect the relations that shape us would not only be undignified but simply cannot be considered as enhancement at all.

Social disruption

Contrary to the optimism shown by the transhumanist paradigm regarding the benefits enhancement interventions would bring to individuals and society, this paradigm considers that individualistic enhancement interventions have reinforced the values from which most social disruption has emerged. It is the values and motivations promoted by a liberal view of individuals which most probably have led human enhancement interventions to threaten and in some instances harm other humans and other living forms.

Enhancement interventions, however, need not be the cause of social disruption. The features of the social paradigm aimed at promoting environments that enable communities and individuals to flourish in harmony are thus less likely to end up in social disruptions than those feared by the biomedical or transhumanist paradigms. For example, given that the idea is not to take over human capacities but rather to supplement them within a framework of social values, scenarios such as posthumans taking over unenhanced humans are unlikely to arise. Moreover, the fact that the paradigm is focused on empowering and enabling individuals allows for more responsive and quick actions to mitigate the impact of social disruptions if they were to arise.

Authenticity

As mentioned in previous chapters there are two interrelated ideas here, cheating and the good life. Connected to cheating, under this paradigm social enhancement is not considered a form of cheating, as interventions are not a shortcut to avoid pain, suffering, effort and hard work. Moreover, the social and relational values underlying social enhancement interventions make it less prone to cheating concerns than the biomedical and transhumanist paradigms with their focus on competitiveness and individualistic values.

Connected to the issue of the good life, it is important to highlight that this paradigm does not start from assumptions that have to do with enhancement interventions being those that make the individual better than well, that enable the individual to experience modes of being that are human-atypical, or that enable the individual to fulfil his or her desires. Rather this paradigm starts from the assumptions that there are various legitimate conceptions of well-being and the good life, and that enhancement cannot be grounded on the betterment of isolated individuals, nor measured by how much technology is involved in our lives. Accordingly, it holds that authenticity as a moral ideal relates to well-being and the good life insofar as it contributes to the discovery, articulation, construction and self-definition of oneself in relation to others and the world. The good life is achieved when we enter into meaningful relations with others, and accept that the relations we have with the world and others shape to a great extent who we are (de Beauvoir, 2011; Sartre, 2012; Taylor, 1991). Thus, under this paradigm, the good life requires more than being authentic to myself as an isolated individual, it requires being authentic to myself as a relational individual. Hence, the good life is only accessible when enhancement interventions contribute to the overall community well-being.

Evolution

This paradigm acknowledges that we are constantly participants in our own evolution; however, this is more from a cultural and social perspective than a biological one. The social paradigm rejects the transhumanist view regarding the idea that *humans* have the tools and knowledge to shape their evolution in ways that biological evolution is fundamentally incapable of producing. While our technology and science can help us to gain understanding and knowledge, it is questionable that this is enough for us to grasp the complex and rich relations that shape us. Moreover, it is questionable that the decision of taking one more step up the biological evolution staircase should be left in the hands of the selected group of individuals comprising the transhumanist group, and under the values they cherish, without a more open discussion and a more relational perspective.

This paradigm also rejects the transhumanist paradigm's view on evolution, which regards evolution as blind, clumsy and unwise because it has not optimised us for certain traits, such as happiness (Bostrom & Sandberg, 2009a). Instead it considers that biological evolution provides most of us with the potential to develop certain traits, and then it is up to us, both as individuals and as society, to optimise them or not. It is also plausible to say that transhumanists are only focusing on the traits they would like evolution to have provided us with, without acknowledging that evolution has already provided us with great features. That is why both the idea of blaming evolution for traits that we humans consider important but that we are not 'optimised' for, and the idea of trying to set standards to measure evolution's outcomes, are undoubtedly poorly grounded.

Furthermore, the fact that we might not agree with the 'wisdom of nature' does not imply that our way of doing things would be a better or a wiser way. It is more likely that we are the ones blinded to the overall picture that is in place when evolutionary processes take one course of action over another, or when such long periods of time are needed to bring change. As Clark has put it:

> [b]iological evolution can select internal coding schemes that look alien and clumsy at first sight but which actually represent quite elegant solutions to the combined problems of serving basic needs and making the most of existing resources.
>
> (Clark, 1998, p. 132)

However, in our case it seems as we have been blinded to this and instead focused on having things our own way, resulting in the need to develop new abilities and skills that were not needed in the past. If this is the case there are not many real grounds to blame evolution for the fact that we might not be evolutionarily optimal to flourish in our ill-created environments. In this regard the social paradigm suggests different venues to foster the traits we want to develop while taking into consideration our relations with the world and others, rather than the individualistic and bodily focused changing interventions suggested by the other two paradigms.

All the above-mentioned points highlight the moral position held by the social paradigm related to human enhancement interventions. Some of the challenges this paradigm has yet to face will be explored in the following section.

The issues

The most daunting challenge of social enhancement is not technological or even economic; it is political and moral. This helps to explain why a social perspective like the one suggested by this paradigm, despite the evidence around the impact our relationships have on our well-being, has not been fully appreciated. Thus, even if we were to consider social enhancement seriously, there are challenges that will need to be addressed, including the following.

Different capacities not just needs and preferences

Within the biomedical and the transhumanist paradigm we mentioned the problem of distinguishing individuals' needs and preferences. Within social enhancement another layer of complexity is added, as it is acknowledged that people have different capacities for adapting and coping with their environments. For instance, some people might be better at handling stressful environments than others. So even if social enhancement acknowledges and explores ways to address these differences, in practice it is not an easy task to choose interventions that will bring overall social benefit. However, in order to better shape future interventions, social enhancement does promote research into how different factors, such as socio-economic position, different vulnerability layers and exposure to the interventions, affect outcomes. While this can be a cumbersome enterprise, it is a feasible and worthwhile option.

The weight of liberal values

This challenge has two different aspects, one connected to political will and community awareness and the second to unnecessary interference. The first aspect deals with the changes that might be needed in order to foster a shift from Western deeply held individualism to one in which community well-being can be prioritised; in other words, how to bring the political will and awareness needed to forward the benefits that social enhancement interventions promise. Today, even though it is well documented that community-oriented interventions bring overall benefits for whole populations and individuals, within the current enhancement discourse it is still individualistic interventions aimed at enhancing individuals that are prioritised. Consider the current pattern in which most of our resources are not directed to the problems of the majority of the world's population (Rabinow & Rose, 2006; Rose, 2007; WHO, 2007). A possible reason for this lack of motivation to support interventions with an overall benefit for communities compared to those focused on individuals is that we seem to undermine the view that we are all vulnerable to changes in our environments, to accidents, and to biological and behavioural factors that might leave us with partial or life impairments at any time. Perhaps if we were to acknowledge our relational nature, people would feel more motivated to promote social enhancement interventions.

The second aspect of this challenge is connected to the idea that for some people social enhancement would constitute an unnecessary interference from different social actors in processes better left to market forces and individual choice. A possible response to this challenge is that we are not in fact discrete and unrelated individuals, but relational and dependent. Thus input from different social actors is not only necessary and desirable, but also unavoidable. Moreover, a focus on individual choice and freedom has led to disintegrated societies. Such kinds of societies, even with biomedical or transhumanist-enhanced individuals, do not have the means to address the global challenges that we face today, challenges that call for collective action. In this regard social enhancement could be a good starting point not only to promote a more relational view of the individual but also to create feasible paths to address global challenges.

Not meaningfully limited

Finally, this paradigm faces the challenge that the notion it gives regarding human enhancement might be interpreted by some people as too broad, rendering almost everything as a form of enhancement. A recent European report on human enhancement (Coenen et al., 2009) suggested that a better understanding of enhancement should be meaningfully limited and exclude such practices as the ordinary use of body-external technological devices, education or physical exercise. While it might be true that such a 'meaningfully limited' understanding of enhancement works well for the biomedical and transhumanist paradigms, which are based on the liberal concept of the individual, it makes no sense under the social paradigm.

A 'meaningfully limited' understanding of human enhancement as suggested in the European report could end up: (1) reinforcing the dichotomous view that the therapy-enhancement debate has carried along since its beginnings; (2) promoting a kind of technological deterministic view on enhancement; (3) reinforcing a view of the individual as separated from the rest of his or her environment; (4) undermining the fact that controversial 'off-label' and 'dual use' of technology do not need to be based on the latest high-technological interventions; and (5) narrowing and restricting the exploration of more meaningful understandings of enhancement. In the case of social enhancement it would be counterintuitive to have an enhancement paradigm based on a relational view and continue to reinforce liberal views about the individual.

It is not that social enhancement is a broad concept of enhancement, but rather that it offers a totally different perspective on the nature of enhancement itself. Limits to what counts or not as enhancement will also need to be discussed, but the nature of these limits will not be the same as those suggested by a liberal view of the individual. As a starting point one can imagine that meaningful limits for *social* enhancement would exclude practices that are individual-bodily based, that are too expensive or high-tech, that are not sustainable interventions, and above all do not recognise the relational nature of individuals.

The value of the social paradigm

Some people might argue that this paradigm is not really about human enhancement. However, considering the relational view of the nature of individuals, the role of the social determinants of health and the way our environments affect our well-being—all key features of the social paradigm—it can be argued that social enhancement is indeed a different and meaningful account of human enhancement worth of consideration. A perspective based on social justice is probably the most meaningful human enhancement alternative, as it promotes actions that include different members of the community, enable and empower individuals as members of communities, and by better grasping the nature of individuals has the potential to bring real improvements for humankind and beyond. Thus, the social enhancement paradigm offers the opportunity to rethink in more fruitful ways different paths to improve ourselves.

The following are some features that make this paradigm a promising alternative to the dominant views around human enhancement.

A more pragmatic approach

A division between those who are blindly enthusiastic about enhancement and those who oppose it has dominated the biopolitics of human enhancement, as portrayed in the transhumanist and the biomedical paradigms respectively. Moreover the distinctions used within the biomedical and transhumanist paradigms (such as therapy versus enhancement or posthuman versus human) have not allowed human enhancement to gain conceptual clearness for policy-makers, the law and its moral discussion. Social enhancement takes a more pragmatic approach, focusing on moving the discussion from the theoretical to the practical (a proactive approach), promoting actions that truly improve the human condition in more ethical ways; and as such it can be seen as an opportunity to bring insight into the human enhancement debate.

This paradigm can also be regarded as a pragmatic approach because it suggests an account of human enhancement in which enhancement interventions do not require necessarily high-technological, radical or individualistic interventions. Thus, while the biomedical and transhumanist paradigms are focused on the latter type of interventions, social enhancement takes on board empirical evidence around the positive effects of environmental interventions, which can be more cost-effective and communally oriented.

A more realistic approach

The social paradigm can be regarded as a more realistic approach because it is not based on blind optimism in technological progress, nor does it accepts that enhancement is necessarily about 'maximisation'. Rather it holds that enhancement is about finding optimal levels. Social enhancement is also a more realistic approach because it acknowledges that enhancement interventions, as any other human intervention in the world, have consequences that go beyond the desired area of change. Following this idea, George Khushf has said that 'an enhancement at one scale and system might simultaneously involve diminishment at another scale or system' (Khushf, 2004, p. 143; c.f. Iuculano & Kadosh, 2013). A clear example of this is the increase in communication technologies, which has helped to connect the global community but has also helped to erode the values of small communities, and has brought more threats to privacy. Another example comes from minimally invasive brain stimulation evidence, where stimulation can facilitate numerical learning while impairing automaticity for the learned material (Iuculano & Kadosh, 2013). In this regard, social enhancement aims to understand more about the complex interactions and long-term consequences of different enhancement interventions, both for the community and for individuals.

A rejection of the concept of the liberal individual

The social paradigm of enhancement, in contrast with the biomedical and the transhumanist paradigms, rejects the concept of the liberal individual and its focus on individualistic interventions, which undermine larger social goods and values of community, care and solidarity. By neglecting the complex and rich relations that shape the human condition, the liberal individual promotes (1) a view where human desires are given the same importance as basic human needs, (2) alienation and segregation rather than membership, as well as (3) fragmentation and social hierarchisation. Such a view of the individual, as Charles Taylor put it, is the root of many 'malaises of modernity' (1991). For example, under the biomedical paradigm, the liberal concept of the individual underlies the belief that what causes disease are isolated entities that we can control, thus individualistic interventions are seen as the path to attain positive health outcomes and human enhancement. Such a belief fails to recognise the network of complex relations in which individuals are embedded (society, social institutions and environment). Once we recognise the relational nature of individuals, we can see why individualistic interventions are not the path to sustained and meaningful human enhancement.

The case of the transhumanist paradigm is no better. For instance, most of their suggested enhancement interventions reflect corporate, military and wealthy able-bodied perspectives. Even in cases where transhumanist enhancement interventions are suggested as a path to help alleviate some of humanity's most pressing challenges, such as poverty, climate change and suffering, these are based on individualistic interventions with the hope that society will somehow benefit from that.

Moreover, although the transhumanist paradigm does not deny that 'deliberate fitting of the environment' is a promising alternative, they consider it just as an '*adjunct* to direct human enhancement for improving human performance and well-being' (Bostrom & Sandberg, 2009b, p. 388 [emphasis added]). The transhumanist paradigm has thus not really been proved to have a real interest in more inclusive and accessible interventions, nor has it shown that the values it promotes are consistent with the values they said to embrace. It is likely that their use of a more inclusive rhetoric is just to 'legitimise' their research agendas.

The social paradigm, in contrast to the biomedical and transhumanist perspectives, breaks free from the liberal individual view and focuses instead on the vast and complex relations that shape human well-being and flourishing, emphasising that individuals are relational beings defined by their relations with others. Thus, the social paradigm gives us a lens through which to see human enhancement possibilities in a new light.

By recognising our relational nature this paradigm also rejects the human prejudice implicit in the biomedical paradigm, as well as the *post*humanism envisioned in the transhumanist paradigm; the latter implying that we need to overcome our human biological limitations to say that we are enhanced. In addition, given that we still have not reached an agreement on what to include within the realm of the human and even less within the realm of the posthuman, the social paradigm tells us that it is not enough to shift to a personhood model, but that acknowledgment of the rich and complex relations that shape us is also needed. More importantly, it tells us that instead of focusing on whether we should stay human or explore realms outside the human realm, we should try harder to live to those standards that we think are worthy of moral consideration in relation to others that might or might not share with us those traits that we value.

Conclusion

Biomedical and transhumanist human enhancement interventions are based on the assumption that individuals are abstract and isolated beings, thus neglecting the social and relational nature of individuals. The social paradigm takes on board the need for a human enhancement framework in which enhancement interventions take into account our relational nature. This chapter started by discussing why the social paradigm is a novel and valuable human enhancement paradigm. The ideas underlying and shaping the social paradigm, namely the social determinants of health and social justice, were also introduced. The main features, views and values of the social paradigm were then laid out. The second part of the chapter examined this paradigm's ethical stand on human enhancement; analysing the set of arguments mentioned for the biomedical and transhumanist paradigms, but from its own perspective. This chapter also pointed out the challenges this paradigm faces as well as the insights it brings to the human enhancement discourse. Thus, despite the limitations the social paradigm might have, it suggests new views about human enhancement that are worth exploring. The next chapter will point out some of the issues have been neglected in the current ethical discussion around human enhancement.

5
To Enhance or Not to Enhance: Looking into Deeper Issues

To enhance or not to enhance?

This chapter questions whether the dilemma around enhancement is deciding whether to enhance or not to enhance ourselves. The current debate around human enhancement has failed to grasp the underlying different value assumptions and power relations behind the different reasons for embracing or rejecting human enhancement. Thus, this chapter aims to illustrate that the real dilemma(s) may rest somewhere other than where the debate is set now.

The first part of the chapter covers ethical issues that are important challenges for all enhancement paradigms, while the second part covers other pressing ethical issues that so far have been neglected within the human enhancement discourse. All of the ethical issues to be presented here are issues that, regardless of the enhancement paradigm we choose, will need to be confronted. However, depending on which paradigm we take as our basis, some of these issues may be more urgent or more pressing than others, such as those involved in interventions that directly affect the brain or are as radical as those suggested by the transhumanist paradigm. One more important consideration to bear in mind is that all the different ethical issues to be discussed here are complex and challenging ones; as such, only an outline of some of the concerns involved can be provided here. Finally the different normative considerations mentioned do not, in and of themselves, support a pro-enhancement or an anti-enhancement stance.

Identity: The core of who we are

Many philosophers call the characteristics that a thing or someone must have, as long as it or she exists, *essential properties*. For any kind of enhancement to be a worthwhile option it has not only to represent a gain in well-being, but 'at bare minimum it must not involve the elimination of any of your essential properties' (Schneider, 2010, p. 224). There are several views of what constitutes these essential properties. One of the most relevant essential properties is that of identity. David Hume makes the point in his *Treatise of Human Nature*, when he remarks that 'of all relations the most universal is that of identity, being common to every being whose existence has any duration' (2004, p. 17).

However, the literature on identity is quite diverse, putting different emphasis on various factors that may be key in forming our identities. One of the most common accounts of identity places memories as its central part. This account starts from the assumption that because our existence takes place in time and because our awareness of it requires self-reflection, our memories should be considered as significantly constitutive of our identities (Locke, 1768; Parfit, 1971; PCB, 2003; Sartre, 2012; Schechtman, 2007). John Locke's theory was probably the first account explicitly linking memory with identity. Locke's memory criterion of personal identity, as it is known, says that an individual (I) at time t_0 is the same individual at time t_1, if (I) has memories of t_0. More specifically, Locke distinguished between being the same person and being the same human (1768). For him, consciousness (as memory of certain past experience) was necessary for being the same person but not for being the same human (the same member of the species human being). Locke's account was not embraced by everybody, leading to the incorporation of different elements into his view of identity. One of these was the psychological continuity criterion, which incorporated forward-looking psychological connections (such as present intention and future action).

Other personal identity accounts focused on the capacity for reasoning or a certain degree of conscious experience, while some other accounts focused on more naturalistic approaches. Some examples of the latter include materialistic accounts in which our identity is based on continuity of our biological material, and biological approaches referring to the continuity of remaining alive biologically speaking rather than just based on psychological continuity (Habermas, 2003). Aristotle held a similar view, arguing that if I were to wish for something for our own good, it would have to be something for myself 'as human being'; otherwise it would be like wishing for my own destruction (Aristotle, n.d.). More recently the idea of somatic identity has become popular. This somaticisation of identity has focused on a particular organ of the body, namely the brain. That is why Rose has suggested that we have become 'neurochemical selves' (2007), changing the way we understand who we are, our thoughts, emotions, preferences and behaviour.

Another account of identity is patternism, supported for instance by people such as Kurzweil and Norbert Wiener, in which identity is based on the individual 'pattern' or patterns of organization of our brains (Kurzweil, 2005; Wiener, 1950).[54] Wiener held that our identities cannot be based on materialism, because '[O]ur tissues change as we live ... We are but whirlpools in a river of ever flowing water. We are not stuff that abides, but patterns that perpetuate themselves' (1950, p. 96). Likewise, Kurzweil said that 'the ordered and chaotic collection of molecules that make up my body and brain' change constantly; therefore it is more accurate to think that we are 'like the pattern that water makes in a stream as it rushes past the rocks in its path. The actual molecules of water change every millisecond, but the pattern persists for hours or even years' (Kurzweil, 2005, p. 383). Thus, according to these views you are essentially the pattern that your brain generates. Some people go as far as to suggest that the brain pattern of an individual can be preserved in a different medium than the body (such as silicon), without altering the neural configuration that constitutes the nature of the person (Bostrom, 2005a; Kurzweil, 2005). This latter view is the presumption held in the transhumanist paradigm, and makes the idea of mind-uploading theoretically possible.

Human identity has different aspects to it, the individual, the collective and the genealogical. Thus, depending on the kind and degree of enhancement intervention, different aspects of human identity can potentially be affected to different degrees. Based on this notion, it would not make sense to ask whether or not identity is affected by a certain enhancement intervention, but rather which aspect is most affected. For example, given that memory is generally considered as 'central to human flourishing', giving us a sense of narrative 'self' that is crucial for experiencing one's life as one's own (PCB, 2003, p. 215), enhancement interventions targeting memories are considered as more problematic to our identities than other types of enhancement interventions. Considering that biomedical and transhumanist enhancement interventions are focused on directly changing the individual, the analysis here will be focused as well on the individual aspect of identity also referred to in the literature as personal identity.

There are two main senses in which personal identity has been generally understood, numerical and narrative identity.

(i) Numerical identity, which is identified by Marya Schechtman (2007) as the reidentification question, deals with the issue of whether an individual at time t is the same person as an individual at another time. Numerical identity then deals with some sort of psychological continuity or continuity of experiential contents over time.

(ii) Narrative identity, which Schechtman refers to as the characterisation question, has to do with the mental states and attitudes, as well as the actions caused by such states, that someone sees as belonging to him or her. Narrative identity involves an individual's self-conception, 'a coherent-self consciousness that extends with past and future stories that we tell about ourselves'. (LeDoux, 2003, p. 20)

Identity and enhancement

Most enhancement interventions, even if only in very subtle ways, affect our identity (Brey, 2009; Coenen et al., 2009; Degrazia, 2005; PCB, 2003; Schneider, 2010). Before proceeding to discuss some of the ways in which our identity can be affected, there are some considerations to be born in mind. First, in what follows, the concept of identity will be used as a set of criteria that must be fulfilled by individuals to persist from one time to another, but must also remain unique with respect to others. Second, the concept of individual-identity will be used as the self-representational system by which individuals define themselves within systems of categories from an early age, which changes and is redefined throughout their lifespan.

Some concerns related to enhancement interventions are that we end up pathologising our identities as syndromes and disorders to be treated (Elliott, 1999), or that we end up disrupting our unique features, thus altering, impairing or even destroying our identities (Coenen et al., 2009; Gordijn, 2006). For example, if we were ever able to upload our minds, Bert Gordijn cautions that this 'could blur the borderline between the self and the cyberthink community' (2006, p. 730). In the case of the biomedical paradigm, therapy presupposes preservation of identity, whereas radical enhancements do not (Fukuyama, 2002; Habermas, 2003; Walker, 2008). In contrast, the transhumanist paradigm does not see any real threat to our identities even in cases of radical enhancement. The stand of the social paradigm is more in line with David DeGrazia's view, which holds that human enhancement per se is not the cause of inauthenticity or violation of inviolable core characteristics (2005). Moreover, it agrees with DeGrazia's view in that as long as we start discussing concerns around identity from different points of view, it is unlikely that we will arrive at sound arguments. In particular, DeGrazia sees numerical identity and narrative identity being used interchangeably within the discussion on enhancement and identity, when in fact they cover different aspects of our identities.

It is highly unlikely that enhancing a person's mental features would create a numerically distinct individual, which means that post-enhancement individuals will still remember life before the intervention without any major alteration to her memories, goals, unique skills, and many important aspects of her personality. However, with novel more sophisticated and precise enhancement interventions this premise is somehow not as stable as it might have been. A clear example, although still a fictional one, would be cases of fusion and brain duplication, which in a way are implied in the suggestions made by transhumanist enhancement interventions such as mind-uploading or brain to brain 'wiring up'. Connected to this point, Susan Schneider has argued that having a particular type of pattern cannot be sufficient for individual identity as 'any sort of uploading case will give rise to a reduplication problem, for uploaded minds can in principle be downloaded again and again' (2010, p. 249), concluding that this would not really be a form of enhancement but rather a form of suicide. This could be understood as a form of identity horizon, beyond which, even if we were somehow connected to the new individual, we would not recognise this version of ourself as ourself; but also as a form in which our mental features could be living in two different substrates simultaneously. In this regard Parfit might be correct in suggesting that identity (whether it is numerical or narrative) should not be our main concern, but rather survival of the individual.

Coming back to DeGrazia's point, even if he is right in that there is no current real threat to numerical identity, we still have the issue that enhancement interventions do affect one's self-conception thereby affecting one's narrative identity. As a matter of fact, this concern is not only common, but also given its plausibility is to some degree a more important one (Schechtman, 2007). For instance, identity crisis can be understood as instances in which there is no coherence between an individual's values and projects and the kind of things they authentically identify with. Moreover, we cannot discard the possibility that certain changes or modifications could alter someone's self-narrative so profoundly that they could result in a different individual altogether. Peter

Whitehouse and colleagues make the point clearer with the following remark:

> Increased memory, new insights, and better reasoning could all lead to new values, new perspectives on one's relationships, and new sources of pleasure and irritation. That does not mean that the enhanced literally will lose their identities and become different people, any more than someone with Alzheimer's does. But in the figurative sense intended by caregivers of people with the disease, it may be that after some point the cognitively enhanced will no longer be recognizable by those who knew them before their enhancement.
> (Whitehouse, Juengst, Mehlman, & Murray, 1997, p. 16)

Another source of concern related to individual's narrative identity arises from the possibility of attaining a posthuman state. For those holding preservation of membership in the human species as a requirement for identity, the attainment of the posthuman state after a given enhancement intervention would be regarded as a threat to narrative identity. Some scholars have argued that this brings issues related to the new being not having consented to being created, and the issue of the individual wishing his or her own destruction (Schneider, 2010; Walker, 2008). Even if we do not invoke the posthuman, we can still imagine having different parts of our biological brains being replaced by artificial parts (for instance because of a flesh-eating disease or after an accident). If we accepted patternism this would not be an issue for identity, as long as the pattern of our brain is preserved, but this would not be true if we held a more naturalistic approach.

Transhumanists have counter-argued that we can see identity changes as a form of 'gradualism' that resembles in a way the changes we incur from infancy to adulthood (Bostrom & Ord, 2006). In order to contrast gradualism with more immediate changes, take Mark Walker's example of the *Homo lamarkian*, a species very closely related to humans.

> As with humans, gestation takes nine months, and when children are born they are indistinguishable (to the naked eye) from human babies. When they are two days old, something amazing (at least to us) happens: the newborns undergo massive amounts of change and then turn into beings that look exactly like adult humans overnight. [...] The Lamarkians are an example of 'punctuated growth' rather than the more familiar gradualism.
> (Walker, 2008, p. 113)

While gradualism seems to leave room for us to do something about the changes we undergo (or at least gives us time to process them), it could also turn out to be like the case of the frog and the bucket of hot water. If a frog is put into a bucket full of room temperature water, which is then gradually heated, the end result for the frog is that it ends up being cooked; however, when the frog is put directly into boiling water it immediately hops out. In a similar way, it could be that with gradual changes we miss the opportunity to notice that something fundamental about us is changing. Thus, by the time we notice the changes, it might be too late to do something about them. The important question to ask here is whether radical changes instantiate different challenges to identity compared to more gradual changes. Other related questions have to do with the number of changes (and their magnitude) that an individual has to undergo in order to alter, in morally relevant ways, his or her identity. At this point these are just rhetorical questions that call for further investigation.

Whether or not a better distinction between different aspects of identity could play a crucial role in moving forward the debate about identity in the moral discussions of enhancement interventions, what is important here is that enhancement interventions do impact in one way or another on identity. Our identities are shaped, sustained and reinforced by our relationship with technology, a relationship that is far from being neutral and uncomplicated, being mostly rather paradoxical, ideological, fantastical and multidimensional (Campbell, O'Driscoll, & Saren, 2010). This relationship gets even more complicated as the times we live in are times in which the boundaries between the human and the technological are blurred and seen as coextensive, mutually defining and co-dependent (Bukatman, 1993). These are also times where the collective is harder to distinguish from the individual self as our identities are dissolved, simulated and reconstructed by our interactions with technology (Hayles, 1999). A clear example is the way the mutability of our self-representations in online environments has resulted in the creation of virtual identities in which our identity need not to be fixed in an unitary notion, but rather can be seen as multiple and changeable (Turkle, 2011; Yee & Bailenson, 2007). Given that these identities are 'an unmistakably doubled articulation in which we find both the end of the subject and a new subjectivity' (Bukatman, 1993, p. 17), they are also regarded as cyborg or posthuman identities (Gray, 1995; Hayles, 1999). This new type of identity brings along changes related to one's representation to others and ourselves in terms of social influence and relationships of power, pleasure, virtuality and reality, all of which deserve serious consideration.

(Self-) knowledge and learning

It is through our relationships that we can appreciate both the uniqueness of each situation but also the similarities of different situations we have been confronted with. The relationships we have with others and our environments shape our views and beliefs about the world and ourselves. Thus, we can see why our relationships with others and the world in general play a key role for (self-) knowledge and learning. While it is plausible to think of some enhancement intervention that enables people to acquire more self-knowledge, for instance by adding emotional valence to a past learning memory or experience, there are cases in which the results might not be so encouraging.

An important aspect for knowledge and learning has to do with the truthfulness of our experiences. Our experiences help us to build a frame of things we believe to be true about ourselves and the world (Bublitz & Merkel, 2009; Elliott, 1998; Taylor, 1991). Therefore, the authenticity of our experiences becomes an important concern. A clear example to illustrate these concerns comes once again from memory enhancement interventions. If in the future these could in a reliable way edit our memories or insert false memories into our minds, it would not only threaten the authenticity of our knowledge about the world and ourselves, but could also lead us to live in falsehood.

Imagine remembering a life, then finding out that one's memories are not really memories of one's life experiences. I can think of conflicting memories, such as remembering being in two different places at the same time the same day, remembering that my car is blue when in fact I am looking at my red car, or having a memory of me and my sister in the beach when I was little but also knowing that I do not have a sister. Such conflicting memories would certainly put into question the truthfulness of our experiences and what we know about the world and ourselves. Some people have argued that given the social nature of remembering there must be a limit 'to how inconsistent our false memories can be' (Liao & Sandberg, 2008, p. 91). This seems to be a reasonable point, unless, of course, we all live in a type of simulation or in a society in which the memories of a large proportion of society have already been altered.

Consider the case of simulations, from computer-based simulations to being immersed in virtual reality scenarios, which are becoming part of our daily living more and more. In some cases our representations of reality (as in the case of virtual experiences) can be so compelling that we risk believing that we have achieved more than we actually have. Philip K. Dick captured this idea succinctly in his speech 'How to build a universe that doesn't fall apart two days later' when he said that fake humans are being created by having fake realities (Dick, 1978). In other words, we might be missing the difference between the real experience and the virtual one (Gordijn, 2006; Turkle, 2011). However, it is still open to debate whether living in virtuality would amount to living in falsehood, or whether there it is of any moral relevance if we truly believe a false memory to be true.

Another possible enhancement intervention that presents a challenge to our knowledge is that of memory deletion. In this case the concern is mainly that by erasing a particular bundle of our past we would not have sufficient self-knowledge or that we would lack some self-referential background information to be able to grasp the kind of person we have become (the person we are now at present). Presumably, the alteration of a few selective memories (such as fear memories) would not have the same consequences as altering a whole chapter of our autobiographical memories. Some people have argued in connection with this that not all memory modifications pose a problem, given that not all memories are crucial for self-knowledge. However, it might be hard to know in advance which memories are important for self-knowledge and which ones are not. A more reasonable line of argument is the one suggesting that given the way our memories are kept in the form of distributed networks of potentiated synapses across different regions and structures of our brains, even if a particular memory was crucial to our self-knowledge it would be highly unlikely that we could achieve a radical alteration of it. Nonetheless, if we were to achieve the kind of memory enhancement interventions suggested by the transhumanist paradigm, it would be fair to raise the concern that such radical ways of altering memory could indeed end up distorting (even if as a side effect) a whole set of other memories, creating a breakdown in the general coherence we see in our lives.

Regarding learning, if people did not need to go through experiences themselves in order to learn from them, then it is plausible to argue that enhancement interventions could indeed challenge common views on learning. Consider, for instance, that we could have a brain implant that enables people to speak Spanish without ever having had any Spanish lessons. For some people the fact that no learning process was involved is not a source of special concern, as they argue that what matters is that the individual has available useful information. However, contra this view, it can be argued that the learning process itself is important because it is what distinguishes us from being mere cogs in a machine. Learning something involves more that the acquisition of new knowledge; thinking otherwise would imply a mechanistic and reductionist account of what it means to be human. Here it is worth pointing out that in the case of simulations at least people are experiencing what it would be like to be in a real situation; thus some form of learning can still be achieved.

Learning is a process that seems to be designed so that only things that are meaningful, important or enjoyable are remembered so that we can use them at a later point. Thus, interventions that were just focused on giving individuals a capacity to remember more things without being able to grasp the meaning or knowing how to use all this information in the relevant context, would more likely impair our ability to learn rather than providing memory enhancement. More importantly, knowledge and learning are shaped by our relational nature. Thus enhancement interventions that do not acknowledge these important aspects are likely to end up impairing more than enhancing.

The moral agent and agency

Closely connected to self-knowledge and learning is the importance of our experiences for enabling the moral agent, particularly if we agree with the idea that agency is shaped by our knowledge about ourselves and the world as well as our learned experiences (Farah, 2010; Greene & Haidt, 2002; Levy, 2007; Liao & Sandberg, 2008).

Some enhancement interventions precisely exploit the links between our experiences and our agency in our best interest. Consider, for example, cases where the harm done to us is so traumatic and damaging to our well-being that we might indeed need to forget the event by weakening, or in the future even deleting, the associated memories. This is the case for many soldiers who 'may not just want to forget that they killed; they may also want to forget how to kill' (Liao & Sandberg, 2008, p. 93), or cases of destructive habits (such as drug abuse), in which even when the person desires to break free of them, it is difficult given their memories (a type of reward-based conditioning) associated with the habits (Bernier, Whitaker, & Morikawa, 2011).

However, there are cases in which enhancement interventions that blur our memories might not be in a person's best interest. Having the capacity to recall past events gives us an opportunity to form ideas of how we should act when confronted with similar situations in the future. Thus when someone is no longer able to recall the things he or she has experienced they might be missing something important in terms of the moral lessons they could gain from their memories. George Santayana captured this idea when he said that 'those who cannot remember the past are condemned to repeat it' (2005). Hence, if someone cannot remember or is not able to grasp and reflect about the moral of his or her previous experiences, it is likely that they will 'keep falling over and over again in the same ditch'. This could end up bringing unnecessary suffering not only to the individual in question, but also to others. Consider the case of someone whose memories of saying bad things to someone they love, things that have hurt that significant other, were erased. It is then possible that the person could end up saying those bad things again, hurting once more that significant individual. This would be the result of not being able to grasp the moral about how those words caused hurt.

In order to show that modifying our memories before we have come to realise the underlying moral of the experience could impair our development as moral agents, Liao and Sandberg (2008) used two examples—the betrayal case (when you are harmed) and the crime case (when you harm others). These cases show that enhancement interventions involving erasure of memories could impair our ability to react in an appropriate moral way (such as forgiving someone after he or she has harmed you or feeling regret after having harmed others). This is one of the main reasons for regarding our capacity to reflect about our experiences and being able to address them as crucial for exercising appropriate agency (Glannon, 2008; Liao & Sandberg, 2008), as well as for respecting ourselves as moral agents (Taylor, 1991). However, that is not to say that all our memories are worth keeping intact: as mentioned above some memories are more integral to our moral selves than others, but the challenge is to distinguish between them.

Some enhancement interventions, such as neural prostheses, also suggest a different configuration of agency, as they are not just a medium but materially reconfigure 'the intersubjective unit body and technology as an intrasubjective entity' (Cartwright & Goldfarb, 2006, p. 138). In the past, most of our mental world was just for ourselves; however, certain enhancement interventions are likely to put at risk the privacy of our mental world. Consider, for instance, the use of mind-reading technologies, which can be used to learn more about our minds and consciousness, but can also be used without the consent of individuals, such as in screening people who are suspected of being terrorists

(Kirsch, 2006). Searching a house is significantly different from going into someone's brain in search of memories and thoughts. Is there a point at which societal benefit could outweigh individual infringement? What about the case of someone who has a memory implant or is using wearable technology with recording capabilities, and is the only witness to a terrible crime? If the police needed to have a look at the recorded information, either stored in the implant or in the device, is it enough to ask the individual for consent, or would the company that made the implant need to be asked as well? These issues could get even more complicated if we consider that others might be connected to our wearable technology input stream.

Moreover, enhancement interventions not only affect our normative status as agents, but in some instances also our capacity to take responsibility for our actions (Hoerl, 1999; Klaming & Haselager, 2010). As Christoph Hoerl points out, 'sensitivity to the fact that certain deeds cannot be made undone is inseparable from the insight that we have to live with the consequences of our past deeds (and of past events in general)' (1999, p. 245). We can also imagine cases in which others might consider important that we remember certain events or features, such as the case of someone who is the only witness to a crime and the only person who could have some important information about the criminals. Do we have a duty not to forget those memories (Levy, 2007; Liao & Sandberg, 2008)? If that were the case, we might have an obligation to use enhancement interventions to assure the preservation of those memories. Certainly, if the memory is affecting the memory-holder, it is hard to see why someone would have a duty to retain a specific memory after this has been stored in a different format outside the brain of the individual (such as in a transcript or recording).

The kind of enhancement interventions available at present might not bring any pressing concerns related to identity, (self-) knowledge, learning and the shaping of the moral agent, but it could also be the case that not much work has been carried out in terms of the impact that non-controversial enhancement interventions have had for individuals and society. It is also likely that as new and more radical ways of enhancement interventions become available some issues could take a different level of urgency. Moreover, even when certain enhancement interventions seem just to be fictional scenarios entertaining the minds of scientists, engineers and philosophers, the fact that there are people working towards achieving these scenarios should give us sufficient grounds to start thinking about the ethical implications. At the same time, there are many different ways in which technology has altered the human condition; some have improved it, others not so much. Thus, in the meantime, more research is needed to investigate whether or not different enhancement interventions brought forward by the use of emergent technologies create ethical issues that are different in kind or degree to those that we have faced in the past with other types of technologies or less technologically driven enhancement interventions.

The other dilemmas

As was mentioned at the beginning of this chapter, there are other issues that have been neglected within the human enhancement discourse. These neglected issues take a particular turn with the use of emergent technologies.

Biopower

The choices we make about our technology have important consequences for the form and quality of human associations and relationships. The use and development of technology is never neutral (Winner, 1977). Power, control and domination have been linked to the development of our technologies and social institutions, whether as a desire to attain power over nature or over ourselves. In this regard, it is possible that Nietzsche was right when he claimed that the will to power is the 'innermost essence of Being' (Nietzsche, 1968). Emergent technologies can be regarded as technologies of power when they are used to master our bodily and mental features, establishing the conduct of individuals and submitting them to certain ends or domination. When emergent technologies are used as a technology of power over life—including health, the body and the mind—it becomes crucial to question the assumptions, values and discourses they bring along.

It was the explosion of diverse and numerous techniques for achieving the control of populations and subjugation of bodies that marked the beginning of an era of 'biopower' (Foucault, 1990; Rabinow & Rose, 2006). Biopower has created the issue of how to define the boundary between private decisions and public regulations, regarding these powers of control one of the most pressing dilemmas for society in addressing human enhancement. According to Foucault (1990), power over life evolved in two basic forms. One focused on the species body— the body permeated with the mechanics of life and as the place of the different biological processes; the other one centred on the body as machine—the optimisation of its capabilities, its integration into systems of efficient and economic controls, the extortion of its forces and its disciplining.

In the core of the human enhancement discourse, one can observe the play of power, truth and values in relation to the subject and to the possibilities of human flourishing, including health, maximisation of quality of life, liberation from pathology and suffering, modification of one's body, satisfaction of needs, happiness, and more and more the discovering of all that one can be. The politics underlying biopower and human enhancement place human existence 'as a living being into question' (Foucault, 1990). The politics underlying these can also entail a kind of enslavement (Lewis, 1943; Nietzsche, 2001), as each new power won by man can be a power over man as well. In connection with this, C. S. Lewis wrote that 'man's final conquest has proved to be the abolition of Man' (1943). Thus, it is possible that in our quest for new and radical forms of enhancement, we are abolishing valuable features of humanity.

For some scholars, such as Robert Goodin (1986), dependence on others creates power relations. Thus given that is unlikely or even desirable that we can eliminate all forms of dependency, the real concern in power relationships is therefore the exploitation of the weak party by the strong one. What is wrong about power, then, is the taking advantage of the implicit dependency positions that it creates (Goodin, 1986). In the case of the social paradigm, this concern about power relationships is not as problematic as it is for the biomedical or transhumanist ones, as the dependency we have with others and our environments is not only acknowledged but also embraced. In the case of the biomedical and the transhumanist paradigms, the individualistic perspective underlying them makes an exploitation of the asymmetry of power likely to occur. The following are some examples of the issues that the power discourses promoted by these two human enhancement paradigms can instantiate:

(1) Social hierarchisation: The types of individualistic interventions promoted by these paradigms have 'acted as a factor of segregation and social hierarchization' (Foucault, 1990, p. 141), in which enhanced individuals somehow belong to the higher social ladder in comparison with unenhanced individuals. The idea of power and dominance centred in life—an idea manifested in the goal of using emergent technologies for controlling our bodies, minds and the whole environment—has also contributed to the promotion of exploitation practices towards the vulnerable.

(2) Well-being and the good life: The view of enhancement held by these paradigms is a view that coerces and manipulates individuals to accept a very distorted and narrow conception of well-being and the good life (Habermas, 2003; Held, 2006). This is a conception that detaches the individual from the complex and rich relationships contributing to his or her well-being.

(3) Objects of consumption: Another issue associated with the biomedical and transhumanist paradigms is that they promote a view in which our bodies along with the traits to be enhanced are seen as objects of 'consumption' or commodities (Hayles, 1999; Miah, 2008). Under such a view our bodies and mental traits are transformed into markers of status and wealth that signify economic success, good taste and social superiority. In this regard, the term 'biovalue' has been used by social scientists to imply the commodification of bodies in a global market where the exchange value is set by new technologies (Nichter, 2008; Rose, 2007), and in most cases used as a negative feature of commodification. The worry associated with these views on commodification and consumption is that it exacerbates the subjugation of our bodies and minds to the preferences of the market. More worrisome is that it fosters a concept of the good life based on the pursuit of rationality, profit, material comfort and convenience, as envisioned by a small group of liberal individuals.

There are other scholars, however, who do not see the commercial character of human enhancement interventions as a negative feature, but rather as a positive one (Miah, 2008). In particular, they argue that these interventions help us to accumulate biocultural capital. The term 'biocultural' is used by Andy Miah as an appeal to recognise the ways in which health has been medicalised and become a feature of cultural politics in healthcare. Extending from Pierre Bourdieu's concept of cultural capital, Miah uses the term to explain human enhancement interventions as acts of consumption, via the consumption of biopolitics. According to Miah, consumption can be an act of differentiation by which people establish their sense of identity, belonging and meaning through the consumption of ideas, products and possibilities.

While it is true that certain acts of differentiation can be used to challenge power discourses based on the exploitation of differences, such as those promoting and embracing biological diversity, it remains to be seen whether or not enhanced individuals once they have acquired more and new powers will still be willing to promote and share interventions that empower the unenhanced and take them away from a 'privileged' position.

Ableism: Towards an ability divide?

Another important but often-neglected concern in the human enhancement debate is that enhancement interventions could eventually impose a one-sided pattern of the development of human capacities and abilities, leading to ableism. According to Gregor Wolbring, ableism is a set of practices, beliefs and processes that produces, based on the favouritism for certain abilities that are projected as essential, a particular kind of understanding of one's body, oneself and one's relationship with others (humans, other species and the environment) (Wolbring, 2005, 2008a).

Based on this definition, there are many possible forms of ableism, including cognition-based ableism, species-typical ableism (in biomedicine) and ableism inherent to a given economic system (such as the ability to consume or the ability to be productive, competitive and efficient) (Wolbring, 2008b). What all these forms of ableism have in common is that any deviation from the favoured ability, real or perceived, is seen as a diminished state of being. Think for example of our modern societies where a certain level of mental capital is seen as essential to be part of the new economic 'playing field' (Lynch, 2005). This is a form of ableism insofar as people see or perceive a certain level of mental capital as needed in order to compete, which could drive people to consider using cognitive enhancers in order to remain competitive. Building from this example, it is not too extreme to say that we have created environments and social conditions that make people desire a certain set of cognitive abilities such as a capacity to concentrate better and for longer hours. When thinking of ableism in this way, we have to ask ourselves about the kind of society and values that we are promoting; in particular, when the option given is 'enhance yourself or fail', where fail can entail losing your job, or being a loser within society.

Different values and visions about enhancement and ourselves are brought forward depending on the human enhancement paradigm that one embraces. Thus, the focus that ableism takes will also vary according to the human enhancement paradigm in place. The biomedical paradigm, for instance, can be linked to a form of ableism that favours 'species typical normative abilities' (Wolbring, 2008b, p. 33). This form of ableism implies that those individuals deviating from species-typical abilities are seen not only as 'less able', but also as in a state of disease, defect or disorder. In the case of the transhumanist paradigm the linked form of ableism favours abilities that are human-atypical, such as hearing frequencies beyond the human limits, having night vision or other types of abilities that are beyond current species-typical limitations. Moreover, building on what has been previously discussed, both the biomedical and the transhumanist paradigms promote a set of values that reinforce competitiveness, medicalisation, discrimination and commodification of the human body (Coenen et al., 2009; Miller, Parker, & Gillinson, 2004; Wolbring, 2008b). In contrast with these two paradigms, the social paradigm encourages and embraces different sets of abilities based on socially oriented aims rather than individualistic ones, preventing the forms of ableism associated with the biomedical and transhumanist paradigms.

It is important to make clear that what is problematic is not the favouring of certain abilities over others, as presumably different people might prefer to develop certain abilities over others according to their own conception of the good life. The favouring of certain abilities over others is problematic based on the set of values underlying the favouring of these abilities, but also when this favouring results in the promotion of values and actions that disrupt society. Favouring certain abilities over others, for instance, can reinforce disablism—a set of beliefs and assumptions (conscious or unconscious) that promote discriminatory, oppressive or abusive behaviour and practices, or unequal treatment of people because of presumed or actual disabilities (Campbell, 2008; Miller et al., 2004).[55] This means that the use of emergent technologies for human enhancement could be problematic when used with the aim of reaching a set of abilities that are the result of created desires (promoted by a consumerist society) or when driven by the unquestioned visions of the good life promoted by certain elites (such as able-bodied wealthy groups).

Another related concern with ableism is that it produces scholarship that contains serious gaps, distortions and omissions regarding the production of normality and abnormality, and re-focuses a certain desired able-bodied lens towards normality and health (Campbell, 2008). A clear example of this is the way in which old people, after losing abilities they used to have when younger (such as being able to move without assistance), are seen as a burden on society. In this frame, our understanding of disability rests more on the tacit understanding of what constitutes being abled rather than on the particularities of people's actual abilities, which undermines the fact that people with different abilities actually do find ways to adapt to their particular conditions. There are cases in which people with an impaired ability for something develop or present enhanced abilities compared to the norm (Silvers, 2008). There is evidence showing that individuals who cannot walk often develop enhanced shoulder strength as a result of using crutches or wheelchairs for mobility. Another interesting case is blind people's abilities compared to sighted people's. While blind people do not have the ability to see the world through their sense of sight, they often develop better verbal memory or better hearing than sighted people. Connected to this, a group of neuroscientists recently reported that blind people can understand spoken language at a rate up to about 22 syllables per second compared to normal-sighted listeners who can comprehend up to 8–10 syllables per second (Hertrich, Dietrich, Moos, Trouvain, & Ackermann, 2009).[56] The study mentioned that this occurs because the brain region sighted people would normally use to respond to visual stimulation is used by blind people to respond to speech. This is an example of how the brain rewires itself in such a way that it compensates for lost senses, and in some cases develops abilities out of what we normally expect in 'normal' individuals. However, there are many factors (such as age) that influence the brain's rewiring capabilities (Hertrich et al., 2009).

Another problem with simply endorsing a view that posits a particular world view that either disfavours or favours dis/able-bodied people, as ableism does, is that it exacerbates issues connected to biopower. Underlying the selection of favoured abilities within ableism, there are various power relationships in place. When these power relationships are not questioned, they become perpetuated standards (through habituation or value shifts), excluding and disabling individuals. This exclusion has the potential to bring to the fore a divide between those with the abilities promoted by the enhancement discourse in place and those that do not have those abilities. This divide can be regarded as an ability divide.[57] Consideration of the various issues brought forward by ableism should make clear the importance of questioning the value of the abilities that some groups of people are trying to promote, as well as the set of abilities that are worth striving for.

An ability divide

According to the capability approach literature, abilities are often put in terms of exercisable functionings, while capabilities are genuine opportunities to realise these functionings (Nussbaum, 2001; Robeyns, 2006; Sen, 1979). Thus, in what follows a distinction between ability and capability is considered, the former being an actual level of development, the latter a potential level of development. However, it is important to keep in mind that ability and capability are interrelated concepts, given that our functionings depend on our available alternatives. In other words, our functionings are a reflection of our capabilities.

Enhancement interventions brought forward by emergent technologies are likely to affect our capabilities by changing their degree, scope, flexibility or duration, or by adding a new capability C (C_1 ... C_{n+1}) to our existing set of capabilities. The latter is sought by transhumanists with their idea of the posthuman, a being who essentially will have a substantially different set of capabilities compared to current humans. Enhancement interventions are also likely to impact our abilities. Here, the concept of an ability divide is used to describe a divide resulting from the use of emergent technologies for human enhancement, in which some people would have improved their species-typical abilities or attained posthuman abilities. A divide like this implies that those who cannot afford or opt not to be enhanced would be regarded as disabled or less worthy than their enhanced counterparts (Cabrera, 2009b; Coenen et al., 2009; Wolbring, 2006).[58]

There is a broad spectrum of human abilities resulting from differences in the *natural lottery*, effort and in some cases even luck. Thus, the problem is not the existence of different abilities; rather the concern with an ability divide, as the one suggested here, is that certain enhancement interventions have the potential to change our abilities in unprecedented ways. These new abilities are not the result of having or lacking a perceptual sense, as it is the case for amputees or blind people. Rather they are the result of directly altering our biological make-up in radical ways, such as enabling atypical human abilities (such as the type of echolocation used by bats) or the ability to download our thoughts and memories directly from our brains into cyber-networks. Such new abilities could have profound consequences for our views about social order and epistemology, as well as communication and even morality.

A communication divide

The last neglected ethical issue regarding human enhancement to be addressed here deals with the impact that new abilities could have on human communication, namely the possibility of a communication divide.

Human communication, in its different forms, matters because it sets a common ground for meeting common basic human needs, such as the need to be recognised by others (Cabrera & Weckert, 2012). It also matters because humans have evolved as social creatures, and it is through communication that we remain social beings. Communication is also an important part of our relational nature, as it enables us to sustain relationships with others. However, successful communication requires a solid base of shared experiences and beliefs (Cabrera & Weckert, 2012). Erik Parens has argued that the ambivalence found in relation to enhancement technologies and practices is the result of different understandings and views about moral ideals held by the critics and proponents of enhancement technologies (2005). While these differences in ethical frameworks can be regarded as a communication problem, these are not the type of concerns that a communication divide, such as the one suggested here, has in mind. However, they clearly illustrate the consequences that different understandings could have for meaningful human communication.

Humans have different forms and degrees of interactions with technology, some of which have changed the nature of human communication (from smoke signals to computer networks). The main premises underlying the idea of a communication divide are first, that certain forms of human communication depend on a shared lifeworld and second, that some forms of human enhancement have the potential to disrupt this shared lifeworld leading to problems in meaningful human communication. However, that is not to say that all human enhancements will lead to communication problems nor even that those that do will lead to insurmountable problems (Cabrera & Weckert, 2012).

Meaningful human communication involves more than transfer of information; it requires a 'shared lifeworld' or common ground, which in turn depends on having similar bodies, perceptual equipment and socially embedded nature. Thus, having common knowledge or common ways of perceiving the world is not enough; instead, as Thomas Nagel (1974) suggested, it is needed shared points of view, which are shaped by values, beliefs, education, and other social and psychological factors. We as humans build up meaning about the world and ourselves as well as creating and maintaining meaningful social relationships, as a result of sharing a set of shared values, abilities, beliefs, knowledge, perceptions, and other social and psychological factors, thus a 'shared lifeworld'.[59]

Accordingly, the more alike our shared lifeworlds are, the easier it is to understand and establish meaningful communication. This helps explain why, in spite of some differences in our shared lifeworlds, we still can attain partial understandings, in which the more subjective viewpoints of individuals play no role in the apprehension of the shared lifeworld. By sharing a human 'lifeworld' we can grasp 'what it is like to be on a rollercoaster'. However when we stop sharing the essential features of a human 'shared lifeworld' the more likely meaningful communication would be impaired or even become unsustainable. Thus, a human 'shared lifeworld', as Nagel has argued, would not be useful when trying to grasp 'what is it like to be a bat' (1974). This idea was also reflected in Wittgenstein's remark: 'if a lion could talk, we wouldn't be able to understand it' (1953, p. 227). We do not have a shared lifeworld with bats nor with lions, thus we cannot really know what it would be like to be a bat or a lion. While I can imagine what it would be like for me, as a human, to be a bat, I cannot imagine what it is like for a bat to be a bat. Some could say that this is a bit of an overstatement, that humans and bats are both mammals, and as such do have quite a lot in common. However, humans and bats lead very different lives, have different body features and at least some different perceptions (Cabrera & Weckert, 2012). A clear example is echolocation, which is used by bats to perceive the external world, by detecting the reflections of their high-frequency shrieks from objects within range. This form of perception has no equivalent in its operation to any human sense. In other words, the neural system that performs echolocation in bats has no homologue in humans. Thus it is not only that 'our own experience provides the basic material for our imagination, whose range is therefore limited' (Nagel, 1974, p. 324), but also that our own human condition limits our access to certain kinds of experiences. A shared lifeworld enables us, then, to secure a horizon 'within which [we as humans] can refer to one and the same objective world' (Habermas, 2000, p. 315). These examples highlight the importance of having a particular kind of perceptual equipment, body and social nature to establish meaning and understanding in communication (Cabrera & Weckert, 2012).

Even among humans, sometimes we find cases where the subjective character of someone's experience is not accessible to us, such as that of a blind person from birth. Up to certain levels of meaning, concepts can be developed to explain the objective facts of the world that each individual lives in (so we could describe for a person who is blind from birth what it would be like to see), regardless of their biological and perceptual differences. In other cases it might be that even when we have the same perceptual channel available we perceive things differently. For instance, you might see the colour of this wall as being red, while someone else might think is closer to orange. However, this kind of subjective difference (in this case of colour perception) is not particularly important as it still allows for understanding. Even in the case of communication between a sighted individual and a blind one, where partial understanding can be attained, there might still be certain aspects that are not accessible to individuals even with very similar biological features.

If this is true, one could imagine future human enhancement interventions which could change so drastically the lifeworld of the enhanced individuals that understanding and meaningful communication could no longer be possible between enhanced and unenhanced individuals. While empirical evidence of the kind of human enhancement interventions needed for such a dramatic lifeworld shift is lacking, it is not far-fetched to suggest that some direct, pervasive and precise enhancement interventions could indeed instantiate such a radical lifeworld shift as to create a communication divide between enhanced and unenhanced. A communication divide in which such enhanced beings would not be able to meaningfully communicate with unenhanced humans (Cabrera & Weckert, 2012).

The following are some enhancement interventions that could potentially disrupt our communication processes:

1. Pharmaceutical interventions: Some stimulant drugs have psychological effects that range from temporal altered thinking process, synaesthesia and altered sense of time and sense of self, to long-term psycho-emotional effects (Cabrera & Weckert, 2012). Not only is meaningful communication questionable when individuals are under the effects of these drugs, but also the experiences of individuals after the effects of the drugs have washed out can be hard to communicate to others. At this point in time very small effects in healthy individuals have been shown from other drugs used for improving episodic and working memory, as well as concentration. That is not to say that in the future other type of pharmaceutical interventions could come to market with the potential to disrupt in more profound ways our 'shared lifeworld'.
2. Brain stimulation techniques such as transcranial magnetic stimulation (TMS) and brain deep stimulation (DBS): When used for human enhancement, these brain stimulation techniques have the potential to trigger emotions or behaviours that are not characteristic of the person in question and that in some cases could produce feelings that go beyond what we as humans have been so far able to relate to (Cabrera & Weckert, 2012).
3. Brain computer/machine interfaces and cognitive/body prostheses: The idea here is that in the future we could have advanced prostheses or brain/machine interfaces that could enable new senses, modes of

perception, or that could affect our bodies such as those providing us with an extra pair of hands, or even a tail.

4. Genetic interventions: If in the future genetic interventions were used to include certain non-human traits in our genomes, it is likely that the set of capacities current humans experience could drastically change. Genetic interventions could also be used to change our bodily features, for instance by genetically engineering different features or numbers of limbs.

5. Mind-uploading: Transhumanists have argued that mind-uploading would enhance human communication, because to date our interactions with others are limited by the 'very low bandwidth of human speech and facial expressions' (Anissimov, 2009). They argue that by directly sharing our memories and parts of our cognitive states with others we could 'engage in a deeper exchange of ideas and emotions' (Anissimov, 2009). Some go so far as to suggest that talking will become a thing of the past. However, even if this could be considered a form of enhancement, it seems to be an instance of a radical departure from shared lifeworlds.

These are some possible examples of the kind of human enhancement interventions that could alter our brains and bodies in unprecedented ways, disrupting the ways in which we perceive, sense and know our environment and ourselves, thereby challenging the possibility of meaningful communication. At this point in time we might not be able to assess the degree and scope of the communication problems human enhancements might bring along, for example the ones mentioned above, but that does not make less important that we start discussing the possibility that such changes might bring.

The communication difficulties highlighted here are of considerable moral importance because of the role human communication has for humans as relational individuals. It is because of the pivotal role human communication plays for the individual and society that a communication divide calls for more seriously ethical deliberation. One ethical concern is that such a divide could bring a kind of hermeneutic crisis (Edgar, 2009; Habermas, 2003). As mentioned above, meaningful communication is easier between people with similar experiences; and only possible between those who share a minimal reference lifeworld. A hermeneutic crisis then would imply the possibility of meaningful communication being eroded. This would include an erosion of the possibility for discussing and deliberating about the implications and applications of new technological interventions as well as the legitimate uses of them. This would also make it impossible to sustain the basis for social interaction, social engagement and cooperation.

Another related ethical concern deals with empathy. Generally speaking, empathy is regarded as the ability to take the perspective of another individual; that is, imaginatively assuming one or more of the other individual's mental states (Goldman, 1993). In everyday language empathy is seen as the ability to put oneself in another's shoes, in order to understand the emotions and feelings of the other. The epistemic form of subjectivity involved here deals with the limits on the understandability or knowability of various facts about conscious experience (Nagel, 1974). That is why only creatures capable of having similar experiences can understand the 'what-it-is-likeness' in the empathetic sense. Thus, while empathy always carries the 'as if' perspective, it acknowledges the differences between the experiences of the individuals in question (they are similar but not identical). Empathy is important because it can play a significant role in motivation and in people's ability to show moral concern for others. Sharing a lifeworld is therefore not only needed for meaningful communication, but also for reaching understanding and agreement of moral propositions (Haidt, 2001), as well as to help us decide whom and what to include within our circle of moral concern. An example of this is the way we are more ready to empathise with the suffering of other humans in comparison with the suffering of other non-human animals. Similarly, we are more ready to empathise with humans with whom we share similar beliefs, values, cultures, political preferences, socio-economic status or lifestyle. That is why it is likely that introducing even more radically different ways of experiencing the world could bring yet more worrisome consequences for meaningful human communication and understanding.

One more ethical consideration has to do with the transcending power we encounter through communication, particularly when we reach an understanding with one another about ourselves and the world (Habermas, 2003). Reaching understanding can be conceived as a mechanism that socialises and individuates in one act. Thus, human communication contains 'the possibility of universal understanding' within the shell of 'the most individual expression' (Habermas, 2000, p. 187). This is one of the most interesting paradoxes of human communication: it bridges and awakens a feeling of unity, while at the same time it also inherently divides, as it entails a process of individuation.

Taking these ethical concerns into consideration and acknowledging the pivotal role human communication has for creating and maintaining meaningful social relationships, then the ways in which human enhancement interventions might impact communication calls for more seriously ethical deliberation.

Conclusion

This chapter has explored some of the issues that have so far been undermined or not considered seriously when discussing human enhancement. The first part of the chapter focused on issues connected to identity, self-knowledge and learning and moral agency. The second part focused on more general issues, including issues of biopower and the possibility of a communication divide. The concerns discussed highlight the idea that problems with enhancement go far beyond those highlighted in the current therapy-enhancement debate, and include issues that come from abuse of power, dominance and new understandings of life itself. This chapter has showed that it is not a trivial matter to promote one human enhancement paradigm over another, as the particular ways in which emergent technologies might be used for enhancement interventions largely depend on the predominant human enhancement paradigm in place. In the next chapter a prioritisation scheme based on the human enhancement paradigm that has the potential to truly benefit and positively impact more lives will be described.

6
A Suggested Approach

Contested paradigms and shifting power relations

In previous chapters, it was argued that different understandings of human enhancement bring forward not only different ways of using emergent technologies, but also that different human enhancement paradigms change the scope and sometimes even the nature of the ethical challenges that each one instantiates. Taking this into consideration, this chapter suggests a plausible approach to the question motivating this book, namely:

> What are the ethical considerations for choosing one kind of enhancement over another, when using emergent technologies?

As a first approximation, most people would agree that pursuing highly important moral goods should take priority over pursuing insignificant ones, and that avoiding serious moral wrongs should take priority over avoiding trivial ones. However, we should not forget that the discourse on human enhancement has brought to light a debate that is both ontologically and epistemologically complex (Rose, 2007). Given the benefits and promises brought up by some forms of human enhancement, a moratorium on enhancement interventions would not be feasible or wise. Likewise, it will be unwise not to take seriously the potential risks and social disruptions involved, even if they look improbable or implausible. Thus, in the middle of this 'conceptual muddle' (Moor & Weckert, 2004) about what enhancement is or should be, or whether it is appropriate or not, there is an urgent need to look for a more concrete, pragmatic and ethical approach to handle human enhancement interventions. It is also desirable to start pointing out and questioning the assumptions and attitudes that have made particular enhancement interventions show up as necessary and meaningful. While it might be true that at a fundamental normative level there is nothing special about human enhancement interventions, and that therefore 'they should be evaluated, sans prejudice and bias, on a case-by-case basis using the same messy criteria that we employ in other areas of practical ethics' (Bostrom & Savulescu, 2009, p. 4), it is also true that some of the differences may be morally relevant (Greely et al., 2008). The kind of human enhancement applications we decide to focus on will shape whether we stay immersed in the biomedical paradigm or whether we shift to a different one. Hence, as a plausible response to the question motivating this book, this last chapter puts forward a suggestion for a paradigm shift in relation to human enhancement interventions. Specifically, the suggestion is a paradigm shift from a pre-dominantly biomedical to a social one. In practice, it is likely that such a paradigm shift will be more like a prioritisation scheme in which social enhancement interventions would be prioritised over biomedical and transhumanist enhancement ones. The suggestion for a paradigm shift can be understood then as a way to address the important areas that other enhancement paradigms have so far neglected but that cannot remain ignored. It is also a way of addressing the misuse of society's medical and human capital resources, the increasing medicalisation of human activities and better understanding the meaning of being healthy or being well. More importantly, it is a shift from an ethics of enhancement based on individualism and autonomy to one based on relatedness and community.

Before continuing, let us recapitulate the main features of each human enhancement paradigm. The biomedical paradigm is mainly focused on individualistic types of interventions. Its main purpose is improvement *of* the individual. Its ultimate goal is the attainment of complete health and well-being, that is to say not just being well, but better than well from the biomedical point of view. While this paradigm focuses on the improvement of individual features and abilities, the improvements it pursues remain within the limits of what is possible for the human species. Finally, the role of emergent technologies within the biomed-ical paradigm is focused on developing or improving medically related interventions that can be used for human enhancement purposes.

The transhumanist paradigm is also focused on individualistic inter-ventions. Its main purpose is improvement *of* the individual and even-tually of the whole human species, as its ultimate goal is posthumanity. Contrary to the biomedical paradigm, this paradigm focuses on the attainment of atypical human features and abilities. The focus of emergent technologies is on radical and often disruptive high-technological interventions, which most of the time are risky and expensive.

The social paradigm is focused on social and community-oriented human enhancement interventions. Its main purpose is improvement *for* society and the community as a way of enhancing individuals. Its ultimate goal is to enable and empower individuals, as relational beings, within communities. The focus of emergent technologies is on environmental or other community-oriented technological interventions that are safe and accessible. One important point to make here is that despite their differences in focus, purpose and goals, all three human enhancement paradigms acknowledge the potential that emergent technologies have for transforming the human condition, as well as the need for thinking about the practical and ethical implications, albeit each paradigm with its particular focus and in its particular way.

Table 6.1 below summarises the main points that characterise each human enhancement paradigm.

It is true that we cannot know a priori whether focusing first on social enhancement interventions will work better than focusing on biomedical or transhumanist ones. That is an empirical question the answer to which we do not yet know. However, considering the different ethical issues that the different human enhancement interventions raise and the global challenges we currently face, we should have enough evidence to identify social enhancement as a reasonable and feasible option to advance human potential. A shift of human enhancement

Table 6.1   Human enhancement paradigms

Paradigm	Focus of enhancement	Improvement [of/for] the individual	Ultimate goal	Main use of emergent technologies
Biomedical	Individual	Of	Be better than well	Medical intervention
Transhumanist	Individual/ species	Of	Techno posthumanity	Disruptive high-technological solution
Social	Society Community	For	Enablement/ empowerment of individuals within communities	Environmental intervention

paradigm may be our only chance to learn the real meaning of being enhanced and to find a more feasible way in which to make sure that we can all enjoy the benefits of enhancement, but more importantly survive it. The two main ethical considerations that are put forward as arguments for the suggested enhancement paradigm shift—a relational view of the individual, social justice, risks—are complex and subject to much debate; thus, here we can only cover an outline of these different ethical considerations.

The social paradigm: Our first priority

One main reason for suggesting social enhancement interventions as our first priority, rather than biomedical or transhumanist ones, is that it departs from the *individual-comes-first* focus inherent in the other two paradigms. Other reasons for trying to separate the concept and use of human enhancement from individual choice include the ambiguity of the therapy-enhancement distinction and the strong transhumanist influence in the human enhancement discourse. By suggesting a shift from the individual to the social level, the intrinsic value of each individual is not negated; rather it is argued that individual well-being can be enhanced through social interventions. Moreover, by focusing on the social level, the intention is to promote cooperation rather than just competitiveness, to ensure that human enhancement interventions do not promote the exacerbation of socio-economic inequalities, nor a view on individuals as abstract and isolated entities. Therefore, a focus on social enhancement interventions promotes a sense of community and relatedness as well as values such as cooperation and solidarity.

This line of argument is by no means new, and many social and community-oriented approaches have argued on the same lines; that is to say, to make the basic structure of society work for the advantage of all. For instance, the communitarian opposition to normative individualism involves the belief that the moral assessment of any institutional order must be based in part on how it treats certain collectives within it, and that such collectives should be given consideration that is not reducible to the consideration given to their individual members (Pogge, 2002). Daniels has also argued that 'the socially controllable factors that promote equal opportunity to develop our given capacities—medical services, traditional public health, and the distribution of the broader social determinants of health—derive special importance from their contribution to protecting opportunities' (2008, p. 21). According to the report by Coenen and colleagues, European approaches to human enhancement, contrary to US approaches, are less focused on individual interventions and instead are aimed at 'finding socially integrated solutions which are largely based on body-external devices' (2009, p. 129). However, this perspective within the human enhancement discourse has not been widely explored.

Morality is, in a way, about bringing harmony between selfish individualistic desires and the social order (Kirmayer, 1988). We are what we are because we are connected to a world, to others, and as such our individual right to be enhanced is connected with our moral obligations towards society and the world in which we live. That is the ambiguity of the human condition: we are individuals but we depend on others. Simone de Beauvoir captures the idea concisely when she stated that 'no existence can be validly fulfilled if it is limited to itself. It appeals to the existence of others' (de Beauvoir, 2011, p. 67). Furthermore, the global challenges we, as humankind, currently face cannot be solved if we keep thinking in individualistic and centralised causes and solutions, undermining the complexity of the systems we are and those we are embedded in.

There are many reasons for not prioritising individualistic enhancement interventions. One is that we cannot expect that enhancement at the individual level will automatically accumulate as an enhancement of society as a whole. This becomes particularly obvious when the political will (from governments and individuals) needed to secure the translation of individual benefits into social ones is absent (Blas & Kurup, 2010; Sarewitz & Karas, 2006). Another reason is that the more high-technological options are available, the less probable it is that people who can afford them will opt for other less technological options, particularly in the kind of consumerist competitive market-driven societies in which most of us live.

In what follows, two main ethical considerations supporting the argument that the views and values held by the social enhancement paradigm should be given priority will be explored. The first is a relational understanding of the individual and the second is social justice.

Relational understandings: From the individual core to its surroundings

The first reason for suggesting a prioritisation of enhancement interventions following social enhancement is that the values and perspectives held by this paradigm are more in tune with what is required for human well-being and flourishing; that is to say, a relational understanding of the individual. Meaningful human enhancement can only be achieved once we acknowledge this aspect of our human condition. We as individuals are part of a community of related individuals, but we are also part of a complex net of interactions and relationships. What this means is that we are more than the abstract isolated individual envisaged by the concept of the liberal individual. We are relational beings not just in relation to others and the world, but also in relation to what constitutes us as individuals. In this section, these two aspects of relatedness and connectivity will be explored.

Neuroselves?

The first aspect of relatedness questions recent and popular views that hold we are not more than the doings of our brain or that we have become neuroselves. Just as there is something seriously wrong with the concept of the liberal individual, there is something seriously wrong with the idea that we are no more than the doings of our brains. That the brain plays a key role in shaping who we are as individuals is not the issue at hand; rather, it is argued that to reduce who we are only to our connections or the patterns in our brain would be missing the point of who we really are. Such a reductionist approach is based on the idea that mind-based concepts can be replaced by neuroscientific concepts (for instance patterns of neural firings), because knowledge of the components of the brain is both sufficient and necessary to understand the mind. This perspective is generally well received in the scientific community, as it allows for building knowledge on tangible and measurable processes rather than on abstract conceptual and meta-representational processing.

Nonetheless, from an ethical point of view, this raises doubts and questions about the genuinely human qualities of our bodies and minds—as it reduces the understanding of who we are to the interaction of atoms and molecules that happen to be part of our brains (Malsch & Nielsen, 2010; Racine, 2010). Therefore, the implications of a strong reductionist stance have been regarded as problematic. This stance, for example, enables new ways of governing ourselves because any mental pathology can be reduced to an 'identifiable, and potentially correctable, error or anomaly' of the organic brain (Rose, 2007, p. 192). Such views have also started to be used to diminish our responsibility, blaming our brains for our actions and behaviours, as if we were indeed controlled by an entity that is not part of us. The Cartesian 'I think, therefore I am' is being replaced by 'it thinks' instead of 'I think' (Borges, 1964).

Moreover, reductionists' views not only make the framing of ethical problems difficult; they also commit a form of 'mereological fallacy', a fallacy grounded in ascribing properties to parts which logically can be ascribed only to the whole (Bennett & Hacker, 2003). Consequently, the relation and interdependence of many components ends up being overlooked. Let us not forget that what happens in our bodies and minds is the result of rich and complex relations.

Fortunately, there are approaches that have tried to respond to some of the reductionist ideas out there, including the extended mind thesis or emergentism. The former, which was discussed in previous chapters, asserts that who we are extends to the set of resources and mechanisms with which we think and interact. This includes the set of tools that we have developed for ourselves, social structures and the environment insofar as they support our mental world (Clark, 2010; Clark & Chalmers, 1998; Haggard & Longo, 2010; Walter, 2009). The latter, emergentism, recognises 'the existence of multiple levels of physical, biological, and social organization that can generate emergent properties of physical, biological, or social systems' (Racine, 2010, p. 195). Emergentism accepts the view that studying the brain can give us insight into the mind, while at the same time it accepts that there are properties not amenable to direct scientific investigation (Varela, Thompson, & Rosch, 1991). Taking these views into consideration, it is plausible to say that the reason for our brain, and consequently our mind, being so important, is not because we can reduce who we are to it, but rather because it is part of a larger system in which a complex and rich number of relational emergent properties occur. That is not to deny the very important role the brain plays in such a system. However, it is in the way it relates to other parts of the nervous system, other systems and other organs, that our brain actually achieves the faculties that we regard as important. In this regard, emergentism gives us a constructive and insightful theoretical and interpretative framework on which to base the first aspect of relatedness and connectivity of individuals.

Relational individuals

This section explores another aspect of our relational nature, the individual as a *whole*, immersed in relations with other individuals and with his or her environment. We as individuals are also embodied, embedded and extended.[60] In the discussion of the social determinants of health, in Chapter 4, the association between an individual's health and well-being and the social conditions in which she lives, works and develops was highlighted (Blas & Kurup, 2010). With this in mind, it was argued that human enhancement cannot be just about how technology will change us, but how we will change technology and our uses of technology in order to truly improve the human condition. Thus, meaningful human enhancement requires responsiveness to the complex and rich relations of which we are part.

Relational views have been explored from a normative standpoint by the ethics of care (Held, 2006; Slote, 2007) and some environmental and sustainability approaches, such as the Gaia hypothesis, a hypothesis that describes 'the biosphere and all of those parts of the earth with which it actively interacts to form the hypothetical new entity with properties that could not be predicted from the sum of its parts' (Lovelock & Margulis, 1974, p. 3). The core thesis of a relational approach, in the realm of individuals, is that we are relational and dependent beings. On the one hand, we are relational insofar as our relations with others— including other beings, our institutions and our environments—define us. On the other hand, we are dependent insofar as we rely on others. For instance, when we are babies (or when we are old or ill) we depend on others to take care of us; we also depend on others to develop and sustain the necessary conditions for living, to learn meaningful ways of communication and to reach deeper relations. Even if we were hermits, we could still say that we were dependent on our environment to provide us the food and shelter we needed to survive. Thus, a relational individual approach rejects the liberal individual view, a view that, as has been explained, overlooks the reality of the human condition by suggesting that we are abstract, independent, autonomous, rational and self-interested beings; a view that has promoted egoism, individualism, competition and the victory of the fittest or wealthiest. Such a view might be appropriate for liberal political and economic theory, but not for the moral realm.

A relational perspective offers us the opportunity to rethink in more fruitful ways how we ought to guide our lives. It also provides us with a framework to see and hear beyond our own individual needs and desires to those from the rich and complex network of relations in which we are embedded. A relational view of the individual fosters social bonds and cooperation, while at the same time paying 'attention to particulars, appreciation of context, narrative understanding, and communication and dialogue in moral deliberation' (Held, 2006, p. 157). Taking these points into consideration, we can see why relational beings are better suited for understanding the realities of global differences as well as the particularities of different situations, groups and cultures. Furthermore, a relational perspective gives us a promising alternative to the dominant moral approaches, which so far are based on the assumption of the liberal individual; thus changing the ways in which moral problems are interpreted and confronted (Held, 2006).

Considering the features that a relational framework brings forward and that the social paradigm is the only enhancement paradigm, from the ones discussed here, that recognises and acts based on a relational framework, it is reasonable to call for social enhancement interventions as our first priority. The biomedical and transhumanist paradigms do not take seriously the fact that improvements within our network of relations (such as other individuals' well-being, as well as our environments and social institutions) not only bring long-term benefits to individual well-being and health, but also that these are shared improvements with our network of relations. Thus, while biomedical and transhumanist enhancement interventions have the potential to improve human health and well-being, they only bring partial improvements for a privileged group of individuals.

Moreover, biomedical and transhumanist enhancement interventions based on a concept of the liberal individual cannot be: (1) sustainable, as they neglect the relation between our limited resources and our never-fulfilled desires; (2) the most equitable option, as technological diffusion tends to be distributed unevenly (at least during the first years), causing unfair and unnecessary burdens of risks and social disruption; and (3) ethical, if they are only concerned with the individual's freedom and often selfish desires, neglecting the way in which such interventions impact the individual's network of relations. Social enhancement offers us an opportunity to explore our relational nature in order to optimally take advantage of our own potentialities and pursue the improvement of our well-being and our human condition in more feasible and ethical ways. Thus, recognising the importance of a relational approach makes social enhancement, in comparison to the biomedical and transhumanist enhancement interventions, a more meaningful path towards enhancing humans.

Justice

The second ethical consideration for the suggested scheme of prioritisation has to do with social justice. Here it is important to consider that social justice and a relational approach go hand in hand; as Neil MacCormick has pointed out, 'justice matters to people who are already in a community with each other' (MacCormick, 1996, p. 309). Without a relational view, justice and its associated values could be blinded by a commitment to individualism. Thus, a comprehensive moral theory, as

Held has argued, will have to include the insights of an ethics of care, and the relational lens that it brings, as well as those from an ethics of justice (Held, 2006, p. 16). A relational framework can help us to keep focused on issues of attentiveness, trust, responsiveness to need, cooperation, fostering of social bonds and cultivating empathic and caring relations, whereas a justice perspective can help us to keep focused on issues related to fairness, equality, freedom and rights (Held, 2006). Hence, even if each has its own sphere of relevant questions, we can definitely benefit from both perspectives.

Issues related to justice have been widely discussed in the human enhancement debate (Buchanan, Brock, Daniels, & Wikler, 2001; Caldera, 2008; Coenen et al., 2009; ETC Group, 2003; Fukuyama, 2002; Habermas, 2003; Lindsay, 2005; Sarewitz & Karas, 2006; Savulescu, 2006; Wolbring, 2006). However, it is not always clear what exactly should be counted as unjust. Should the fact that certain enhancement interventions are not accessible to everyone be counted as unjust? Many systems in place, for example private schools or certain medical interventions (such as organ transplants), are systems to which not everyone has equal access.

Should it then be counted as unjust that certain kinds of enhancement interventions benefit certain individuals but not others? Surely bad luck in the natural lottery does not necessarily impose any obligations on others (Levy, 2007); but if we accept this line of thought, then selfish desires should also not impose any obligations on others. Moreover, even when enhancement interventions are equally accessible to individuals, it cannot be assumed that every enhancement intervention will have the same outcome for different individuals. The effects enhancement might have are embedded in complex situations, and as such are dependent on the social and contextual background in place (Rose, 2007). Take, for example, the case of alcohol. The same volume of alcohol will have quite different behavioural, emotional and cognitive consequences depending on whether it is taken on a solitary and melancholic evening at home, at a social event (such as a party or a football match), or in the controlled setting of a psychological experiment.

Should it be counted as unjust when an individual enhancement causes societal disruption? But what exactly is meant by a societal disruption? Societal disruption could include feelings of discomfort from seeing other individuals with certain non-typical capacities, but it could also be about enhanced individuals discriminating against non-enhanced individuals.

For our purpose here, we do not have to settle these issues as long as we agree with the idea that enhancement interventions alone will not lead to unjust results. The worst outcomes of enhancement, in terms of justice, do not come from enhancement technologies not being accessible to everyone, rather from the fact that their benefits as promoters of equality of opportunity might not be available to everyone (Allhoff, 2005; Chatterjee, 2004; Daniels, 2008; Farah et al., 2004; Greely et al., 2008).

Even when social justice arguments are not new or particular to emergent technologies, there are two main reasons that have motivated a focus on social justice regarding emergent technologies-based human enhancement interventions. Firstly, a skewed distribution of human enhancement interventions is likely not only to exacerbate feelings of discrimination between the enhanced and the unenhanced, but also to promote depression, anxiety and feelings of disempowerment. It can be said that in societies where people might feel coerced to have enhancement interventions for themselves or for their children in order to be able to stand a chance in the competitive societies they live in, feelings of disempowerment and anxiety are likely to arise. More importantly, a skewed distribution of these kinds of enhancement interventions has the potential to instantiate far greater and meaningful divides than previous ones, such as an ability divide or a communication divide.

The second reason to focus on social justice is that justice is among the most important virtues of social institutions (Rawls, 1971), which are essential to ensure the proper development, distribution and legislation of emergent technologies-based human enhancement interventions. In addition, social institutions are also essential to broaden the discussion concerning human enhancement beyond academia and techno-progressive groups of people.

As mentioned throughout Chapter 4, social enhancement is committed to the goals and underlying values of social justice, such as developing and promoting sustainable and widely accessible applications. In addition to this, the fact that the social enhancement paradigm is motivated by social justice could help us ensure: (1) that the benefits promised by human enhancement are available to everyone rather than just to a privileged group of people; (2) that the interventions are responsive to the basic needs of the community and the relational individual; and (3) that equal levels of opportunity are reached. On the other hand, this could also help us prevent harmful kinds of ableism, power struggles (such as those underlying biopower) or divides such as the ability or communication divide discussed earlier. All these points would seem hard to attain within the biomedical and transhumanist enhancement paradigms, as they depart from a rather different idea of what should count as just with regard to social justice. Taking into consideration all these reasons, together with the presumption that social justice is of key importance for human well-being, makes the suggestion to prioritise social enhancement a reasonable one.

The issues discussed in this section should give us enough moral reasons to consider prioritising the use of emergent technologies for human enhancement applications as suggested by the social enhancement paradigm. The use of emergent technologies for human enhancement applications following a social enhancement paradigm can be regarded, then, as a better and more definitive ethical option compared with the other enhancement paradigms. First of all, because the use of emergent technologies under the social enhancement paradigm is focused on interventions that are relatively inexpensive, easy and not necessarily high-tech but are 'likely to be able to achieve comparatively big results' (Bostrom and Roache, 2011). Second, the use of emergent technologies under the social paradigm acknowledges the differences between various social actors and the necessity to more inclusive approaches towards individual and community well-being. Thus, enhancement interventions under a social paradigm are more likely to enable the conditions needed for people to live the lives they value and the conditions in which individuals and communities can be empowered. Overall, improvements in our environments may be effective and widely acceptable enhancement options.

Moreover, social enhancement brings social justice and a view of the individual as related and interdependent to the foreground of enhancement interventions, both of which are needed in order to shift the focus of our attention and actions towards addressing urgent global challenges (such as public health or climate change). Most, if not all, of the global challenges we face at present call for more social action and social awareness. Furthermore, as long as the benefits of enhancement interventions are not fairly distributed and accompanied by a shift of the motivations and values from the current dominant ones, enhancement interventions might turn out to be worse for us than no enhancement at all.[61]

Thus, for the moment, shaping our environments to mitigate the effects of our limitations or to constrain harms is likely to be more effective and preferable than biomedical or transhumanist enhancement interventions. However, taking into consideration that there are certain realities of life that the social enhancement paradigm might not be able to tackle right away, the suggested second best option in the scheme of prioritisation would be biomedical enhancement interventions.

The biomedical paradigm: Our second option

There are several reasons for suggesting the biomedical paradigm as a second option in the suggested prioritisation scheme. First, research on the social determinants of health has shown that health can be achieved through cheaper and more reliable ways than those suggested by conventional biomedical interventions. That is to say, biomedical interventions are not always the most cost-effective interventions available. For example, development of new drugs is an extremely daunting task, with high economic costs, long development time (usually 10–15 years) and just a few cases actually making it to final clinical use. Second, it is questionable that using technology for individualised and high-tech health systems really fits with overall social needs, in particular if we consider that currently there is a substantial number of people without access to basic health care. Third, most biomedical interventions are based on the presumption of control over the body rather than wisdom about the body (McKenny, 1997). Thus, even if biomedical enhancement interventions acting as pre-symptomatic could avoid the need for scarce and expensive specialists and equipment later on, they could end up promoting the individual's dependency on biomedical interventions to deal with body imbalances, rather than on the individual's inherent healing mechanisms. Likewise, there is evidence suggesting that the misuse and unnecessary use of certain biomedical interventions has promoted the development of stronger forms of pathogens (O'Neill, 2014).

While the points just discussed already give us enough reasons to question the use of this paradigm as our first option, there are two main ethical considerations for suggesting biomedical enhancement as our second option after social enhancement. The first has to do with evidence related to biomedical interventions and their impact on mental health and burden of disease. The second consideration has to do with issues related to safety and risks.

Empirical evidence: Mental health and burden of disease

Health has become not only a priority in relation to policy-making and funding decisions. Moreover, insofar as different ethical theories consider lessening human affliction and suffering of significant moral importance, health is also regarded as fitting in with the goals of the main ethical theories. It is of special moral importance because it 'contributes to the range of exercisable or effective opportunities open to us' (Daniels, 2008, p. 3). The moral importance of health is reflected by the fact that it is even protected as a human right by the Universal Declaration of Human Rights.[62]

A very important aspect of health is mental health. There are more than 1,000 different brain and nervous system-related illnesses, injuries and disorders—such as Alzheimer's and other dementias, addiction, depression, anxiety disorders, brain tumours, epilepsy, multiple sclerosis, Parkinson's disease, psychotic disorders, stroke and trauma—some of which bring particular challenges to health care and society. As a group these conditions (which can alter thinking, mood and/or behaviour) are among the 10 leading causes of disability worldwide, result in more hospitalisations than any other disease group, including cancer and heart disease (WHO, 2008), and constitute 12% of total deaths globally (WHO, 2007), putting a social and financial burden on society. Another reason to focus on mental conditions is because these affect around a billion people worldwide, from all geographical regions and socio-economic status, across all developmental stages and with an overall similar prevalence rate for men and women (WHO, 2007, 2008). However, according to almost all studies women show a higher prevalence of depression, anxiety and eating disorders; whereas men show higher rates of substance abuse, autism and attention-deficit/hyperactivity disorder (WHO, 2008).

To give an idea of the numbers, the World Health Organization estimates that 24 million people suffer from Alzheimer's and other dementias; 15 million from addiction; and 154 million from depression (WHO, 2008). These can be the result of genetic risk factors, injury, abnormalities in the structure of the brain (such as tumours), substance abuse, exposure to heavy metals, metabolic disorders, dietary deficiencies or stress, to mention just some sources. Moreover, these conditions will, most likely, increase in the coming years (WHO, 2008) as a result of demographic changes (such as an increase in lifespan) and modern living conditions.[63] This will be a challenge for care systems and societies worldwide, as the social and economic impacts resulting from these conditions affect individuals, families and the community (Kirkwood, Bond, May, McKeith, & Teh, 2008; WHO, 2007).

The biomedical paradigm, with its focus on medical interventions, could help address some of these challenges by providing preventive interventions or by restoring lost or impaired functioning as a result of these conditions. In this regard the use of emergent technologies can help us to create tools and produce new knowledge that, at least in principle, will be aimed at improving the way we deal with these conditions. Thus, regardless of the controversial therapy-enhancement distinction that the biomedical paradigm holds, advances in emergent technologies can make a big difference for both health and well-being.

In order to emphasise the potential that the biomedical paradigm has for helping us to address brain and mental conditions, this section has pointed out some of the empirical evidence related to these conditions. However, this is not the only reason for considering the biomedical paradigm as our second-best option. There are considerations of balancing risks and benefits that are important to take into account.

Risks: Possible harms versus promised benefits

One major ethical consideration for all enhancement paradigms is the benefits-risk ratio. The great advances that we might be able to accomplish using emergent technologies for human enhancement will not come without their own challenges: as our technological options grow, so does the incalculability of their unknown and unintended consequences. Generally speaking, risk involves a relationship between the probability of something happening and its severity (Allhoff, 2009a; Beck, 1992). Thus, a technological society as ours, where there is high uncertainty about consequences, can be thought of as a risk society (Beck, 1992). At the same time, our views about technology, medicine and the nature of individuals in a way blind us to the real social, cultural and political risks involved in certain enhancement interventions.

A related concern deals with issues about indeterminacy—an inevitable gap between the limited experimental conditions and reality—and ambiguity between different frames of meaning and interpretations in relation to the risks involved (Ingeborg & Dalmo, 2007). Issues about indeterminacy and ambiguity play a key role insofar as the debate on human enhancement is driven by the public's perception of risk, rather than the actual risks themselves (Bostrom & Ord, 2006). Connected to this, Bostrom and Ord have argued that when we are presented with a double epistemic predicament such as radical uncertainty about risk evaluation and prediction, our judgements about such matters 'are not based exclusively on hard evidence or rigorous statistical inference but rely also—crucially and unavoidably—on subjective, intuitive judgment' (2006, p. 657). Indeterminacy and ambiguity are also important when we encounter extreme attitudes towards risks, for instance, when people stop thinking about risk and thus become passive about it, or when they become obsessed with trying to prevent all risk. The latter is a worrisome position because a risk-free society would be a static society (Bostrom & Ord, 2006).

Arguments connected to benefits and risks are common across the three different human enhancement paradigms. The biomedical paradigm is characterised for its concerns with risks, arguing that the possible risks exceed the possible benefits; the transhumanist for being optimistic about the benefits, arguing that the benefits exceed the possible risks; and the social paradigm being a middle-ground view between the biomedical and the transhumanist paradigms.

Among the common arguments related to risk, we can mention the following. One has to do with the idea that the great benefits of enhancement interventions are often interwoven with greater risks. Consider for instance nanoparticles' ability to cross the blood brain barrier—a feature that would enable more targeted and presumably more effective interventions. While this opens new possibilities for treatment it also opens the possibility of more dramatic and irreversible damage. Similarly, while enhancement interventions might improve certain aspects of our well-being or human features, they are likely to end up impairing others (Iuculano & Kadosh, 2013). For instance, a certain memory enhancement intervention might indeed improve our working memory but it might impair other types of memory or other mental features. Likewise, enhancement interventions could elicit harms that individually are practically negligible, but when they aggregate (either in the individual or in society) the net result is a larger and much substantial harm (Lin & Allhoff, 2008b). Take for instance, the possible pressure or burden on society if many individuals decided to undertake life extension, which could include overpopulation, and allocation of job opportunities and resources.

Another argument has to do with the fact that the use of emergent technologies for enhancement interventions affects us in more direct and pervasive ways. Related to this is the issue that the promises of improvement are oftentimes hyped or inflated, which can make people undertake certain enhancement interventions when in fact there are no real benefits or where the risks are higher than those reported. Studies have shown, for instance, that cognitive enhancement drugs are generally not as effective for those with above average capability compared to cases in which they are used to attain average capability performance (Farah, 2010).

Considering the features of different enhancement interventions according to their respective paradigms, it seems reasonable to say that transhumanist enhancements interventions are the most risky, followed by biomedical, and the least risky (at least in principle) those suggested by the social paradigm. One reason for arguing that social enhancement is the least risky option comes from the assumption that interventions that are not directly changing our biological nature are less risky than those that do. Similarly, many of the possible risks involved with social enhancement interventions are known, and there are already effective and acceptable risk reduction strategies available (WHO, 2002). Thus, overall there are more risk reduction strategies available and fewer unknown direct risks involved in social enhancement interventions compared with biomedical and transhumanist enhancement interventions. Another reason for thinking that social enhancement risks are somehow less worrisome is because of the relational framework underlying this paradigm, which enables a better understanding of how risks arise and how to handle them better. This includes an understanding of how the way in which risks are perceived changes according to different contextual and cultural factors.

In terms of the biomedical paradigm, there are a couple of reasons for suggesting that the risks involved locate it in an intermediate position between the transhumanist and the social paradigm. First, even though therapies have to go through a fairly thorough process of research and testing before they are released to the market, they can still pose many known and unknown health-adverse side effects. Often, medical treatments carry the risk of producing anticipated side effects—such as headaches or nausea—that we may consent to in order to be rid of a given condition, but they also bring unanticipated side effects, some of them deadly or life-lasting. Now imagine the case of biomedical enhancement interventions, for which there are not yet clinical trials or control groups providing evidence about their safety, efficacy or possible side effects. This means that the evidence needed to make an informed decision between risk and benefits is lacking. Moreover, therapies with known high risks are generally accepted because they bring a larger benefit (in most cases these therapies are life-saving), which is not the case for most biomedical enhancement interventions. At this point in time, enhancement interventions benefits do not seem to outweigh their risks.

Another reason for the biomedical paradigm to be considered our second-best option is because this paradigm presupposes that disease can be fought independently of the individual's different system interactions and even environmental influences. Considering the relational nature of individuals, it can be argued that such a reductionist view about individuals' healing processes brings forward unnecessary risks and unknowable risks. There are not only known and unknown risks, but there could also be 'known unknowns' and even 'known unknowables'. The latter types of risk are those most associated with creating uncertainty.

The mechanism used by the biomedical paradigm to deal with the uncertainty involved in enhancement interventions tends to be slowing down research and development of interventions, holding off the undertaking of certain enhancement interventions or banning certain paths of research. In the literature the motivation underlying these approaches is described under the concept of the precautionary principle (Allhoff, 2009a). Here it is important to highlight that in contrast with the precautionary principle approach embraced by the biomedical paradigm, the social and transhumanist paradigms take a proactionary view, in which human enhancement interventions are promoted, albeit under the values and assumptions of their particular frameworks.

This section has discussed the empirical evidence we have around mental health and the different risks and risk mechanisms involved in relation to emergent technologies under different human enhancement paradigms. The points presented here together with the different features of each enhancement paradigm give us strong moral reasons to position the biomedical paradigm neither as our first priority but also not as the last one in our suggested human enhancement priority scheme.

The transhumanist paradigm: Not our main priority

Within pluralistic societies there will always be people who think they should be left to do what they desire, including undertaking very radical types of human enhancement. However, there is evidence that the majority of people still believe that certain enhancement interventions (such as biomedical or transhumanist) are not needed (Cabrera, Fitz, & Reiner, 2014). If we take into consideration the fact that we live in societies under restricted resources and poor regulatory schemes, we can see why many people do not consider these types of enhancement interventions as priorities (Hamlett, Cobb, & Guston, 2008). All these points give us enough moral ground to argue that the transhumanist paradigm should be considered last. This section will explore further the main reasons for this position.

Even in a liberal society in which freedom to intervene in our bodies and minds is regarded as unquestionable, as long as we do not harm others, there is still room for the view that certain type of interventions, inasmuch as they disrupt the advancement of a just society, should be left for later. This does not mean that we should put a ban on transhumanist radical ideas; rather the argument here is that there are other things that need to be taken care of first, such as addressing urgent global problems (public health or climate change, for example) and keep working towards the achievement of a basic set of human capabilities for all humans. There are two main reasons that will be presented here for considering the transhumanist paradigm as our last priority. The first is connected to the goals and values it holds, the second to the unprecedented types of risk it could instantiate.

Goals and values

The values and goals cherished by the transhumanist paradigm are focused mostly on the needs and wants of a privileged group of individuals. Consider, for instance, the goal of using advances in emergent technologies to *fight* ageing. On the one hand, it is questionable that ageing should be considered as a disease or something to be ended. On the other hand, there are more pressing worries faced by humanity than continuing to prolong lifespan, such as sanitation or proper living conditions for individuals living on less than two dollars a day. While it is true that within the transhumanist paradigm there are references to the idea of ending suffering and disease, the type of interventions suggested by them do not seem to address conditions such as hunger, diarrhoea or mental impairment owing to a lack of micro-nutrients during early development. Another example highlighting that the values and needs held by the transhumanist paradigm are those of a privileged group of individuals is their goal to reverse-engineer the brain. This goal requires an incredible large amount of energy and processing power.[64] The fact that there are other more urgent uses for energy and economic resources provides us with one more reason for arguing that the use of emergent technologies for transhumanist enhancement interventions should not be a priority (at least not for now).

Another reason comes from the value put by the transhumanist paradigm on our biological aspect. This paradigm considers human biological nature as limiting the exploration of other modes of being that are worth pursuing. This idea is used to argue for technology taking hold and even replacing most of our biologically related processes and features. However, let us remember that many of our functions are based on the 'use it or lose it' principle. Evidence from monkeys who are raised from birth to six months of age with one eyelid closed, for example, shows that the animals permanently lose useful vision in that eye because of diminished use. If we extend this idea to enhancement interventions that aim to take control over our mental and bodily capabilities, then we should question whether in the long run the result of using our biological features less and less is truly an enhancement. In other words, it is not clear why we should prioritise interventions that will likely make us lose functionality rather than interventions that would promote optimisation of our biological given functions.

One more point has to do with this paradigm's view of what is valuable and good in someone's life. According to the transhumanist paradigm, it is the attainment of different posthuman capacities that is considered valuable for individuals, a posthuman capacity being 'one which is much more excellent than that which any current human could achieve unaided by new technology' (Bostrom, 2009, p. 11). Indefinite health span, improved intellectual and bodily faculties, as well as higher emotional capacities are among the posthuman capacities that transhumanists regard as most valuable. Recapitulating what was mentioned in previous chapters, these are valuable features under the following assumptions. First, most people place a very high value on their continued existence in a healthy state; that is to say, most people would choose the path of 'prolonged life, health, and youthful vigour over the default route of aging, disease, and death' (Bostrom, 2007, p. 7; Kurzweil, 2005). Second, most people have a desire to improve their cognitive and bodily functioning, including remembering more things, concentrating better and understanding more. Third, posthuman modes of being enable the exploration of entirely new sensibilities, perceptions and modalities. Connected to this last point, Bostrom has argued that it is likely that there are 'new psychological states and emotions that our species has not evolved the neurological machinery to experience, and some of these sensibilities might be ones we would recognize as extremely valuable if we became acquainted with them' (Bostrom, 2009, p. 11). Finally, most people seek to educate and ennoble their sentiments, to build a character and reduce negative feelings in their lives (Bostrom, 2009).

There are many ways in which these assumptions held by the transhumanist paradigm can be counter-argued, but here we will explore only two. The first has to do with the idea that enhancing just our biology or psychological features would not guarantee a more valuable state of being. As was mentioned when discussing the main motivations for a social enhancement paradigm, our relationship with our natural and social environment, and even the relationship we have with our technology, contributes greatly to our well-being (Clark, 1998; Kirmayer,

1988). Thus, it is debatable that we can attain more value in our life by neglecting the rich and complex network of relationships that contribute to our well-being and that ultimately help to define who we are. This gives us enough evidence to support social enhancement, at least for the time being, as a priority over the biomedical and transhumanist paradigms. The social paradigm is the only enhancement paradigm that holds an alternative view of the individual to the liberal one, regarding individuals as relational beings. That is why it has been argued that the use of emergent technologies under a liberal individual view— where our functioning is reduced to the working of a type of biological machinery isolated from environmental and social factors, where bodies and individuals are seen as abstract and isolated agents—cannot ensure the achievement of any meaningful improvements for the human condition. In this regard, a social enhancement approach to the use of emergent technologies for human enhancement interventions not only rejects a view in which we try to match or improve our mental and bodily capacities to that of our technologies, but it also considers that in order to truly enhance the human condition, the way our physical and social environments shape us and how we shape them cannot be overlooked.

The second point that counterargues the transhumanist assumptions mentioned above has to do with the promoted values. The systematic devaluation of notions of relatedness, interdependence, trust, solidarity, empathy and mutual concern occurring under the transhumanist paradigm encourage egoism, non-interference, self-determination, competition and the victory of the fittest. This set of values is not likely to enable enhancement interventions where everyone can benefit or at least not make their situation worst off (such as being discriminated against or seen as disabled). Acknowledgement of these two points will hopefully make clearer why this paradigm should not be taken as our first priority.

Unprecedented risks

All human interventions face issues of unpredictability of outcomes, short- and long-term side effects for different individuals and society, and issues about being used with a different purpose from the one it was intended or tested for. Thus, as mentioned above, risks are inherent in all three paradigms; albeit each paradigm deals with risks differently. It has been pointed out that transhumanist enhancement interventions are the most risky compared to the other two paradigms. For instance, whereas in the biomedical paradigm it is rational to weight risk against benefits, in order to choose interventions without unnecessary risks (WHO, 2002), within the transhumanist paradigm the kind of risks that are rational to undertake follow a different rationale. Accordingly, in the case of transhumanist enhancement interventions, the question stops being whether there is enough scientific evidence of benefit or individual betterment to make risks worth undertaking, but whether human desires should triumph over the kind of risks we promote. Considering the new capabilities for manipulating life that emergent technologies put in our hands and how susceptible and interconnected our well-being is to our environments and to others, the more important it is to question the transhumanist's position around risks.

The fact that we have worked towards improving our technologies over time has not been enough to stop the crash or collapse of technological systems, or the negative impacts of our technological developments. If the idea is to directly manipulate our mental capital and biological features via invasive and radical technological interventions, we have to ask ourselves what happens when things go wrong. What happens if there is a technological failure or if there are biocompatibility issues? Or what happens when someone hacks our implant? These issues might seem too abstract and not that relevant for the current ethical discussion, but they remind us that these are complex issues that require deep thinking and analysis. It also reminds us that we cannot keep reducing our overall well-being to what happens to us as isolated individuals. This is why relational perspectives are so important. In this regard, as Eric Racine and Cynthia Forlini (2010) have argued, further research is needed to better understand the ethics of enhancement at a social level, and to determine which paradigm or which combination of paradigms reflects the views of the different 'stakeholders' involved.

Conclusion

This chapter started by contrasting the three human enhancement paradigms discussed in this book. It summarised the main features and pointed out the main differences among them. Building on the different ethical issues discussed throughout previous chapters as well as the outline of some specific ethical considerations, a prioritisation scheme was suggested in which social enhancement should be considered as our first priority, followed by biomedical enhancement and finally transhumanist enhancement. More specifically, this chapter argued that given the rich and complex relationships shaping our well-being and health, together with proper consideration of the kind of global challenges we currently face, a social enhancement perspective offers us with a more pragmatic, reasonable and ethical way to use emergent technologies for human enhancement. It was also discussed that given empirical evidence in relation to mental health and the types of risk involved in different enhancement interventions, the biomedical paradigm offers us a suitable option for using emergent technologies after we have considered social enhancement interventions. Finally, it was maintained that consideration of the specific values and goals, risks and costs underlying the use of emergent technologies locates the transhumanist paradigm as our last priority, at least for the time being.

7
Conclusions

The previous chapter sketched a prioritisation scheme to help us decide, based on the different ethical stands and issues each human enhancement paradigm brings forward, how to better use emergent technologies for human enhancement interventions. This final chapter revisits the main argument of this book, provides a brief summary of the major issues discussed in each chapter and ends up with some recommendations for moving forward discussion and actions in this area.

Re-thinking Human Enhancement builds on and contributes to work in the enhancement debate by forcing us to consider uncomfortable questions about the human condition and about values that we generally take for granted. In light of the new capabilities that emergent technologies put in our hands to manipulate and control matter and our bodies and minds, humanity's emotional and intellectual capacity to use emergent technologies well was questioned. The main argument advanced here is that the different set of values and visions underlying our understanding of human enhancement can have a significant impact on the way emergent technologies are used with the goal of improving the human condition. These sets of values and visions can also have an impact on different societal views of science, technology and their future, different social expectations and desires, and different values and concepts about the meaning and nature of being human. With this in mind, three different human enhancement paradigms were explored: the biomedical, the transhumanist and the social. For each of them their main motivations, underlying values and assumptions, their target and purpose of intervention, as well as the role emergent technologies are likely to play, were examined. After discussion of the ethical issues that each paradigm presents us with, a suggestion was put forward to prioritise the use of emergent technologies, in which the social enhancement paradigm should be considered as our first priority over the biomedical and transhumanist paradigms.

The first chapter introduced the main subject to be discussed, namely human enhancement, providing an overview of the general discussion around the topic. It was stated that the all-too-human desire to improve ourselves and to overcome our limitations, together with the new tools emergent technologies put in our hands, are two main reasons for the increasing attention on human enhancement. This chapter also described the main problem with the current debate about enhancement, as well as the main aim and scope of the book. The second part of this introductory chapter provided an overview of the technologies that are likely to play a central role in the future of human enhancement. Nanotechnology and neurotechnology were the two technologies that the book set to cover more broadly, but a brief introduction to other technologies, including information technology, robotics and artificial intelligence, was also provided in this part of the chapter. The final part of the chapter described why the technologies just introduced are likely to play an important role in the pursuit of enhancement— from their unique capabilities to the separate ethical domains they have instantiated.

The second chapter dealt with the current predominant view on human enhancement, the biomedical paradigm. The chapter started with an overview of the concepts of health, disease and the goals of medicine, all of which have shaped this paradigm. It then set out the main features of the paradigm: (1) enhancement as contrasting with therapy; (2) a focus on the individual as an isolated and abstract being; (3) the purpose of enhancement interventions being that of directly changing the body or mental features of the individual beyond better than well but still constrained with species-typical features; and finally (4) the role of emergent technologies for biomedical enhancement interventions. This chapter also discussed the main arguments that have been provided to support the precautionary (bioconservatist) position that this paradigm holds in relation to the use of emergent technologies for enhancement interventions. The underlying issues of such an understanding of human enhancement were also discussed, including the idea that there is not a clear line between therapy and enhancement. The chapter ended by examining the value that this paradigm can have, for instance, being practical in terms of policy-making when dealing with enhancement.

The third chapter dealt with the transhumanist paradigm. It started by laying out the main ideas framing this human enhancement paradigm, namely transhumanism and posthumanism. It then set out the main features of the paradigm: (1) enhancement as the next logical evolutionary step and even as a moral duty; (2) a focus on the individual as an isolated and abstract being; (3) the purpose of enhancement interventions being that of directly changing the body or mental features of the individual beyond species-typical features and ultimately the attainment of the posthuman stage; and finally (4) the role of emergent technologies for radical, high-technology, often costly and risky transhumanist enhancement interventions. This chapter also discussed the main arguments that have been provided to support the pro-enhancement position this paradigm holds. Among the ethical issues surrounding this enhancement paradigm, holding a too mechanistic and reductionist account of what contributes to human well-being was discussed. This chapter ended by suggesting the promotion of biological diversity and further exploration of the posthuman concept as valuable contributions of this human enhancement paradigm.

Chapter 4 explored the social enhancement paradigm. This chapter started by laying out the main ideas framing this paradigm, namely a view of the individual as relational and dependent, the social determinants of health and a commitment to social justice. It was maintained that social enhancement, with the new perspectives it brings to the nature of individuals and the role of new technologies for improving the human condition, is a genuine enhancement alternative worthy of consideration. The main features of the paradigm were set out: (1) a view of enhancement in which improvement of our human condition cannot be achieved by neglecting the rich and complex relations shaping our well-being; (2) the purpose of enhancement interventions being that of changing the environments that disabled us to perform in our optimal levels, both physically and mentally; and (3) a focus on using emergent technologies for creating fair, sustainable, accessible environments that enable us as individuals and as members of communities to develop and make optimal use of our given capabilities. The main arguments supporting the use of emergent technologies for enhancement interventions (social position) were stated. The possibility of not being considered a meaningful limited understanding of human enhancement was mentioned as one of the main issues this paradigm has to deal with. The chapter ended by examining the value of this paradigm: its promotion of a more pragmatic and ethical approach towards human enhancement and the promotion of a different set of values compared to those promoted by the concept of the liberal individual.

With the purpose of helping to clarify the main differences between the three suggested human enhancement paradigms, each of the chapters dealing with a specific paradigm presented examples of how emergent technologies can be used according to their respective enhancement paradigm. Within the biomedical paradigm, for instance, emergent technologies are mostly used towards pharmacological and genetic interventions, but also have recently started to focus on neural interfaces and brain stimulation. The role of emergent technologies under the transhumanist paradigm is mostly oriented towards developing smart drugs, nanobots and the merging of humans with machines via more sophisticated neural interfaces. Finally it was stated that the role of emergent technologies under a social enhancement paradigm is focused on developing environmental interventions, such as smart sensors, textiles and buildings.

The next chapter discussed ethical issues that all three human enhancement paradigms have to dealt with, as well as ethical issues that have been neglected or that have not been as widely discussed but are nevertheless of significant importance. Connected to the former type of issues, identity, self-knowledge, learning and moral agency were examined; whereas for the latter, biopower, ableism and new divides (ability divide and communication divide) were discussed.

Building on these different ethical issues, Chapter 6 synthesised the different human enhancement paradigms and presented an argument for prioritising the use of emergent technologies for human enhancement. Based on a relational view about the individual and as a matter of social justice, it was argued that when using emergent technologies for human enhancement interventions our first priority should be the social enhancement paradigm. Then, based on the idea that in some cases directly changing an individual's body or mental functioning could be integral for mitigating or preventing biological impairment and even death, biomedical enhancement interventions were proposed as a second option. This left the use of emergent technologies for transhumanist enhancement interventions as our last option. That is to say, transhumanist enhancement interventions should be considered only after we have covered global priorities, achieved the enjoyment of the fundamental set of human capabilities for most humans, and society has agreed to move forward with this type of enhancement (not just a particular privileged group of people).

While this prioritisation scheme is a reasonable first step, we should not underestimate the stakes or the difficulty of the conceptual and the practical work—both moral and political—ahead. To conclude, here are some recommendations and points to be considered for moving forward human enhancement from discussion to fruitful actions.

First of all, enhancement *per se* is not something we should fear or embrace without being critical about it; rather we should question the values and visions supporting the particular understandings around it. Thus, before starting any serious debate about the propriety of enhancement, we really need to ask ourselves which are the core values, qualities and abilities that we want our societies to be governed by.

Second, the effects of human enhancement for humankind are not likely to be uniform, playing out differently on the global stage depending on acceptance, investment and a variety of other decisions; thus whether biomedical or transhumanist forms of enhancement are more problematic or useful than 'social' enhancement remains to be seen. However, regardless of the normative framework around human enhancement that we decide to promote, an approach more attentive to the relational nature of individuals and to the kind of social practices that particular enhancement interventions promote remains an urgent need within the human enhancement discourse. At the end of the day, we should not be looking for an absolute superior approach, but rather for an approach that can deliver the most feasible, sustainable and ethical criteria for human enhancement.

Third, emergent technologies are neither good nor bad: they can be used for different applications and bring different results when introduced into different contexts or under different set of values and visions. Hence the prioritisation scheme suggested here is just a starting point to further stimulate more necessary work on the area of human enhancement, not only more discussion on short-or mid-range ethical questions, but also covering more comprehensive issues, such as our core values and its implications for the understanding of ethics, public health and the meaning of being human as individuals and as members of rich and complex communities. Particularly we need a discussion that is more sympathetic to the multicultural and pluralistic world in which we live. Further research into the factors behind people's acceptance or rejection of enhancement is also needed, together with research into the kind of enhancement interventions pursued, and even how enhancement is understood from different perspectives, such as different cultural background, socio-economic status, age or gender.

Finally, the purpose of analysing three different human enhancement paradigms mainly showed that the different visions and values underlying our understanding of human enhancement have important consequences for the way we view ourselves as humans, as individuals and the role we ascribe to our technologies in helping us to improve our capabilities and the human condition. This will hopefully stimulate a broader social and academic discourse on the topic, but more importantly move the debate to sustainable policies and actions. The real benefits of human enhancement will be revealed in our creativity to understand human improvement and our capacity to use as best as we can our technologies for that purpose. Much work remains to be done on the philosophy and ethics of enhancement.

Notes

1 Introduction to the Enhancement Debate

1. Here, it is taken that the human condition is not the same as human nature. The term 'human condition' is not only a simple collection of basic features characterising humans, nor implies an essence that remains identical from birth to death; rather, the term implies something dynamic and in a process of continuous development and negotiation. For more on this see Arendt, *The Human Condition* (1998), and Carnevale and Battaglia, 'A Reflexive Approach to Human Enhancement' (2014).
2. Consider for example that nowadays people sometimes talk about themselves in terms that in the past we only used for our creations, such as updating and even upgrading ourselves.
3. Here and in what follows, the term 'individualistic' will be used as a view resulting from holding the concept of the liberal individual.
4. According to the report it is a 'non-medical' typology since there is no specific definition of health involved. However, it is plausible to argue that by including the term 'therapeutical', they are indirectly referring back to a biomedical-based definition.
5. In the rest of this book, science and technology will be referred to interchangeably, even though it is acknowledged that there are differences between them. For the purpose of this work it is enough to understand that both science and technology have affected and shaped the material condition of our lives as well as the way we understand ourselves as humans and social beings.
6. Here, the term 'nanoscale' is used to refer to the nanometre scale.
7. In his book *Engines of Creation* (1986), Eric Drexler uses interchangeably the term 'molecular technology' with nanotechnology, as a result of nanotechnology's power to manipulate molecules.
8. Available at: http://www.iso.org/iso/iso_technical_committee?commid=38 1983.
9. For arguments about the ambiguity of the US Nanotechnology National Initiative definition of nanotechnology see Fritz Allhoff's paper 'On the Autonomy and Justification of Nanoethics' (2007).
10. Buckyball is the short form of buckminsterfullerene, a molecule cluster of 60 carbon atoms with a geodesic structure.
11. Bottom-up self-assembly occurs when structures are built up by adding atoms or molecules in the right sequence under the right conditions.
12. For specific companies and product names, the Project on Emerging Nanotechnologies (PEN) keeps an inventory of products on the market that uses nanotechnology. Available at: http://www.nanotechproject.org/.
13. The NNTI definition was first suggested by the Neurotechnology Industry Organization (NIO) in their Neurotechnology Public Policy Tour, and later introduced to both houses of the US Congress in the NNTI Act.

14. Lynch has described in this field neuroceuticals as neuromodulators that, contrary to current psycho pharmaceuticals, have high efficacy and negligible side effects (2004). Among the neuroceuticals the envisions are cogniceuticals (focused on learning, decision-making, memory processes and attention), emoticeuticals (influence awareness, moods, motivation and feelings) and sensoceuticals (focused on the capacity of our senses).
15. These different fields are based on the NIO classification (2007) and Lynch's classification (2005).
16. The experts referred to Kant's distinction between a physiological anthropology and a pragmatic anthropology. The former is based on a scientific understanding and manipulation of the brain, whereas the latter is based on knowledge about the social sphere, the world and human behaviour.
17. These features are a summary from those mentioned in the HLEG Report (Nordmann, 2004).
18. Quote from the first Spider-Man story: 'Amazing Fantasy #15'. Later quoted in the *Spiderman 2* movie, 2002.

2 The Biomedical Paradigm

19. Epistemological dualism differentiates between the knowledge and power of the known (the physician) and the known (the patient).
20. Pathophysiology refers here to the concatenation of biological events that distinguishes function from malfunction.
21. Not everyone believes that disease has a normative component that calls for treatment. Christopher Boorse (1977), for instance, has argued that disease description can be value neutral; however, others such as George Agich (1983) and myself have questioned this position.
22. This relationship became known as the patient–physician relationship in biomedicine.
23. Some scholars have argued that medicine has no essential domain of practice, hence a coherent distinction between medical and non-medical interventions can never be drawn in the first place (Engelhardt, 1976).
24. Higher integration density refers to the higher number of channels in a system. Higher numbers of channels are also needed to handle larger amounts of raw data without increasing the size, power consumption or compromising long-term stability of the system.
25. A good example is the development of *in vivo* lab-on-a-chip (LOC) systems. LOC systems promise to be less invasive (owing to miniaturization), to require less sample material than previous diagnostics tests and to provide real-time monitoring of biometric indicators (Tomellini, Faure, & Panzer, 2006).
26. Selgelid has used the example of baldness to make the point that therapy and enhancement can be seen as the extremes of a continuum.

3 The Transhumanist Paradigm

27. For an analysis of Nietzsche's work and transhumanism see Sorgner, 'Nietzsche, the Overhuman, and Transhumanism' (2009). For a critique

of Nietzsche's work being used to support the transhumanist agenda see Bainbridge, 'Burglarizing Nietzsche's Tomb' (2010).

28. The WTA was an international non-profit organisation whose activities ranged between Internet discussion, development of documents, representation in the media, organizing the annual conference 'TransVision' and publication of the online *Journal of Evolution and Technology*. The association as such was ended in 2009 and transformed into the humanity+ organization. See http://humanityplus.org/.

29. These are just some examples from the different forms of transhumanism mentioned in the FAQ list of transhumanism and Anissimov 'Transhumanist Sects' (2006, December 19).

30. http://www.spectrum.ieee.org/singularity.

31. Bostrom uses general central capacity to refer to: health span, cognition and emotion. However, he also acknowledges that by limiting his list he does 'not mean to imply that no other capacities are of fundamental importance to human or posthuman beings' (Bostrom, 2009).

32. Some people prefer using the term 'post-biological' (Kurzweil, 2005) to stress the technological dominance over the biological, without having to go into the murky area of what it means to be human in the first place.

33. Here by biopolitics it is understood all specific contestations and strategies over problematisations of collective human vitality, morbidity and mortality; over regimes of authority, the forms of knowledge and practices of intervention that are legitimate, desirable and efficacious, that have the potential for developing new dimensions of political opinion (Rabinow & Rose, 2006).

34. Clark and Chalmers in their paper suggested certain criteria to be met by non-biological candidates for inclusion into an individual's cognitive system, such as (1) the resource be typically invoked and reliable available; (2) that the retrieved information is 'more-or-less automatically endorsed' just as we generally do with the information we retrieved from our biological brains; (3) that the 'information contained in the resource should be easily accessible as and when required' (Clark & Chalmers, 1998).

35. David Kirsh and Paul Maglio (1994) calculated that for the Tetris game, the physical rotation of a shape through 90 degrees takes about 800–1,200 milliseconds by mental rotation (unchanged brain), physical rotation can be performed in as few as 100 milliseconds, plus about 200 milliseconds to select the button. Here it is worth mentioning that neither Clark and Chalmers or Kirsh and Maglio argued for preference towards a neuro-implant.

36. The term 'cyberspace' is used here to refer to a produced space that moves the subject's relation to reality inside a 'non-space', in this case a digital format common to our ICTs.

37. There are, of course, many versions of transhumanism; as such, some of the criticisms made here may not apply to all versions. These critics are focused on the particular view of transhumanism put forward in the transhumanist paradigm section.

38. The biomedical paradigm also has deterministic views but with an emphasis on biological determinism (such as genetic determinism or neurodeterminism).

39. Bostrom has argued (2005c) that transhumanism does not imply technological optimism; however this is not the general ethos of the transhumanist movement. Bostrom has (recently) tried to use a more value-neutral term such as 'technological development' instead of 'progress' in order to avoid the problems arising from a focus on progress.
40. Philippe Verdoux pointed out the historical argument as another argument to consider; however, here the historical and the over-generalisation arguments are considered together (Verdoux, 2009).
41. By technogenic it is emphasised not only that something is caused by technology but also that it is anthropogenic (Verdoux, 2009).
42. Bostrom explicitly leaves aside the question of whether all pairs of possible lives have commensurable value.
43. I have argued elsewhere that depending on which view we hold and promote about the posthuman, it could either turn out to be our worst nightmare (being the twilight of human race as we know it) or an enhancement of our humanness (Cabrera, 2009b).

4 The Social Paradigm

44. Particularly if we take into account the sensationalist language that the media has adopted to talk about human enhancement.
45. As Cascio nicely captures it 'Transcendence is social, not solitary' (Cascio, 2006, p. 89).
46. Discussing in depth the ethical arguments and different positions involved around these topics could be the topic of entire research projects; as such, here only an outline of these issues can be offered.
47. Available at: http://www-personal.umd.umich.edu/~delittle/nussbaum.htm (accessed October 16, 2014).
48. A similar view has been put forward in Bostrom and Roache (2011), arguing that smart policy should encourage interventions that maximise benefits for society and need not be expensive, controversial, risky or difficult.
49. The reason for being cautious about fully embracing the idea that our minds are extended is because even when external symbols can be integrated in our cognitive processes they are still passive in as much as we still need a nervous system and brain to process and decode the information that our senses capture from our environments (such as cues and prompts).
50. The suggestion here is nothing like geo-engineering, which seems to be a solution offered by some transhumanists when thinking about not so individual-based enhancement interventions.
51. See https://www.kickstarter.com/projects/unchartedplay/soccket-the-energy-harnessing-soccer-ball (accessed November 2, 2014).
52. We should not confuse chemical interventions with pharmaceutical interventions, which are a subgroup of chemical interventions. This paradigm is focused mainly on environmental chemical-detector devices, such as in paint, food, plastics and textiles.
53. For instance, it is generally acknowledged that in order to make sustained changes in people's behaviour, sustained changes in the environment are needed.

5 To Enhance or Not to Enhance: Looking into Deeper Issues

54. Some people have referred to this view as implying computational continuity rather than psychological continuity.
55. For some people, the term 'ableism' is used interchangeably with the term 'disablism'. However, some scholars have argued that the terms render quite radically different understandings of the status of disability in relation to the norm (Campbell, 2008).
56. Humans cannot talk that fast, which means that the researchers had to synthesise fast speech, which for most of us would just sound like noise.
57. Some scholars have argued that if we are referring to the kind of abilities that will be attained after enhancing ourselves, we could address the divide as an enhancement divide instead (Lin & Allhoff, 2006). However, given the different enhancement paradigms covered here, the concept of an ability divide will be used.
58. Elsewhere I have argued (Cabrera, 2009b) that the ideal would be that enhancement technologies were used to help us shift from disability to enhanced ability for everyone.
59. Habermas has also used the concept 'shared lifeworld', referring to a set of skills, competencies, knowledge and perceptions used to negotiate our way through everyday life. See Habermas, *The Theory of Communicative Action: Lifeworld and System: A Critique of Functionalist Reason* (1985).

6 A Suggested Approach

60. Whether the self can be constructed outside a biological body and we can live as patterns of information in a virtual environment (e.g. a computer-based environment) is still an open question to be further investigated, but one to which a relational approach would say no.
61. Ingmar Persson and Julian Savulescu have argued on similar lines, but they focused on the idea that even if cognitive enhancement interventions were available to everyone, they will need to be accompanied by 'moral enhancement that extends to all' (Persson & Savulescu, 2008, p. 166); because as long as 'there is a minority which is morally corrupt', this type of intervention can result in far worst results than no enhancement at all (p. 163).
62. Available at: http://www.un.org/en/documents/udhr/index.shtml.
63. For instance, it is considered that by 2030 depression will have become the second highest cause of burden of disease.
64. Evidence from the Swiss-funded project Blue Brain has shown that just emulating one neocortical column of a rodent brain needs an incredible amount of energy.

References

Agich, G. J. (1983). Disease and Value: A Rejection of the Value-Neutrality Thesis. *Theoretical Medicine and Bioethics, 4*(1), 27–41.

Akerman, M. (2009, February 3). Global Learning Device on Social Determinants of Health and Public Policy Making. Pan American Health Organization.

Allhoff, F. (2005). Germ-Line Genetic Enhancement and Rawlsian Primary Goods. *Kennedy Institute of Ethics Journal, 15*(1), 39–56.

Allhoff, F. (2007). On the Autonomy and Justification of Nanoethics. *Nanoethics, 1*(3), 185–210.

Allhoff, F. (2009a). Risk, Precaution, and Emerging Technologies. *Studies in Ethics, Law, and Technology, 3*(2), art. 2, 1–27.

Allhoff, F. (2009b). The Coming Era of Nanomedicine. *The American Journal of Bioethics, 9*(10), 3–11.

Allhoff, F., Lin, P., Moor, J., & Weckert, J. (Eds.) (2007). *Nanoethics: The Ethical and Social Implications of Nanotechnology* (Hoboken, NJ: Wiley).

Allhoff, F., Lin, P., & Moore, D. (2009). *What Is Nanotechnology and Why Does It Matter* (West Sussex: John Wiley & Sons).

Allhoff, F., Lin, P., Moor, J., & Weckert, J. (2010). Ethics of Human Enhancement: 25 Questions & Answers. *Studies in Ethics, Law, and Technology, 4*(1), art. 4.

Alpert, S. (2008). Neuroethics and Nanoethics: Do We Risk Ethical Myopia? *Neuroethics, 1*(1), 55–68.

Amato, I. (1999). *Nanotechnology: Shaping the World Atom by Atom* (Washington, DC: National Science and Technology Council).

Anissimov, M. (2006, December 19). Transhumanist Sects. *Accelerating the Future Blog*, http://acceleratingfuture.com/michael/blog/2009/01/what-are-the-benefits-of-mind-uploading.

Anissimov, M. (2009). What Are the Benefits of Mind Uploading. *LifeBoat Foundation*, http://lifeboat.com/ex/benefits.of.mind.uploading.

Antal, A., Nitsche, M. A., & Kincses, T. Z. (2004). Facilitation of Visuomotor Learning by Transcranial Direct Current Stimulation of the Motor and Extrastriate Visual Areas in Humans. *European Journal of Neuroscience, 19*, 2888–2892.

Arendt, H. (1998). *The Human Condition*, 2nd edn (Chicago, IL: University of Chicago Press).

Aristotle. (n.d.). *Nicomachean Ethics*. Translated by W. D. Ross. The Internet Classic Archive, http://classics.mit.edu//Aristotle/nicomachaen.html.

Arnall, A. H. (2003, July). *Future Technologies, Today's Choices* (London: Greenpeace Environmental Trust).

Bach-y-Rita, P., & Kaczmarek, K. A. (1998). Form Perception with a 49-Point Electrotactile Stimulus Array on the Tongue: A Technical Note. *Journal of Rehabilitation Research Development, 35*(4), 427–430.

Bach-y-Rita, P., & Kercel, S. W. (2003). Sensory Substitution and the Human–Machine Interface. *Trends in Cognitive Sciences, 7*(12), 541–546.

Bainbridge, W. S. (2010). Burglarizing Nietzsche's Tomb. *Journal of Evolution and Technology, 21*(1), 37–54.

Bainbridge, W. S., & Roco, M. C. (Eds.) (2005). *Managing Nano-Bio-Info-Cogno Innovations: Converging Technologies in Society* (Dordrecht: Springer Netherlands).

Baker, L. (2011, November 2). Tracking Air Pollution Exposure Using Smart Phones: New Study. *Medical News Today*, http://www.medicalnewstoday.com/articles/216026.php.

Banks, D. (1998). Neurotechnology. *Engineering Science and Education Journal, 7*(3), 135–144.

Bayertz, K. (2003). Human Nature: How Normative Might It Be? *The Journal of Medicine and Philosophy, 28*(2), 131–150.

Baylis, F., Kenny, N. P., & Sherwin, S. (2008). A Relational Account of Public Health Ethics. *Public Health Ethics, 1*(3), 196–209.

Beaglehole, R., Bonita, R., Horton, R., Adams, O., & McKee, M. (2004). Public Health in the New Era: Improving Health through Collective Action. *The Lancet, 363*(9426), 2084–2086.

Bear, M. F., Connors, B. W., & Paradiso, M. A. (2007). *Neuroscience—Exploring the Brain* (Baltimore, MD: Lippincott Williams & Wilkins).

Beck, U. (1992). *Risk Society: Towards a New Modernity* (London: SAGE).

Bennett, M. R., & Hacker, P. M. S. (2003). *Philosophical Foundations of Neuroscience* (Oxford: Blackwell Publishing).

Berger, F., Gevers, S., Siep, L., & Weltring, K.-M. (2008). Ethical, Legal and Social Aspects of Brain-Implants Using Nano-Scale Materials and Techniques. *Nanoethics, 2*(3), 241–249.

Berger, T. W., Ahuja, A., Courellis, S. H., Deadwyler, S. A., Erinjippurath, G., Gerhardt, G. A., et al. (2005). Restoring Lost Cognitive Function. *Engineering in Medicine and Biology Magazine, IEEE, 24*(5), 30–44.

Berne, R. W. (2006). *Nanotalk: Conversations with Scientists and Engineers about Ethics, Meaning and Belief in the Development of Nanotechnology* (Mahwah, NJ: CRC Press).

Bernier, B. E., Whitaker, L. R., & Morikawa, H. (2011). Previous Ethanol Experience Enhances Synaptic Plasticity of NMDA Receptors in the Ventral Tegmental Area. *Journal of Neuroscience, 31*(14), 5205–5212.

Bijker, W. E., Hughes, T. P., Pinch, T., & Douglas, D. G. (2012). *The Social Construction of Technological Systems: New Directions in the History of Technology*, Anniversary edn (Cambridge, MA: MIT Press).

Bioethics. (2012). *Emerging Biotechnologies: Technology, Choice and the Public Good* (London: Nuffield Council on Bioethics).

Bioethics. (2013). *Novel Neurotechnologies: Intervening the Brain* (London: Nuffield Council on Bioethics).

Bircher, J. (2005). Towards a Dynamic Definition of Health and Disease. *Medicine, Health Care and Philosophy, 8*(3), 335–341.

Birnbacher, D. (2009). Posthumanity, Transhumanism and Human Nature. In A. M. Cutter, B. Gordijn, G. E. Marchant, A. Pompidou, & R. Chadwick (Eds.), *Medical Enhancement and Posthumanity* (Dordrecht: Springer Netherlands).

Blackman, S. (2009, November 1). Promises, Promises. *The Scientist*, http://www.the-scientist.com/?articles.view/articleNo/27746/title/Promises–Promises/.

Blas, E., & Kurup, A. S. (2010). *Equity, Social Determinants and Public Health Programmes* (Geneva: World Health Organization).

Boenink, M. (2009). Molecular Medicine and Concepts of Disease: The Ethical Value of a Conceptual Analysis of Emerging Biomedical Technologies. *Medicine, Health Care and Philosophy, 13*(1), 11–23.

Boorse, C. (1975). On the Distinction between Disease and Illness. *Philosophy & Public Affairs, 5*(1), 49–68.

Boorse, C. (1977). Health as a Theoretical Concept. *Philosophy of Science, 44*(4), 542–573.

Borges, J. L. (1964). *Labyrinths* (London: Penguin Books).

Bostrom, N. (2002). Existential Risks. *Journal of Evolution and Technology, 9*(1), 1–31.

Bostrom, N. (2005a). A History of Transhumanist Thought. *Journal of Evolution and Technology, 14*(1), 1–25.

Bostrom, N. (2005b). In Defense of Posthuman Dignity. *Bioethics, 19,* 202–214.

Bostrom, N. (2005c). Transhumanist Values. *Review of Contemporary Philosophy, 4,* 87–101.

Bostrom, N. (2007). Dignity and Enhancement. In The Presidents' Council on Bioethics, *Human Dignity and Bioethics: Essays Commissioned by the President's Council on Bioethics* (Washington, DC: US Government Printing Office).

Bostrom, N. (2008). Letter from Utopia. *Journal of Evolution and Technology, 19*(1), 67–72.

Bostrom, N. (2009). Why I Want to Be a Posthuman When I Grow Up. In A. M. Cutter, B. Gordijn, G. E. Marchant, A. Pompidou, & R. Chadwick (Eds.), *Medical Enhancement and Posthumanity* (Dordrecht: Springer Netherlands).

Bostrom, N., & Ord, T. (2006). The Reversal Test: Eliminating Status Quo Bias in Applied Ethics. *Ethics, 116*(4), 656–679.

Bostrom, N., & Roache, R. (2007). Ethical Issues in Human Enhancement. In J. Ryberg, T. S. Petersen, & C. Wolf (Eds.), *New Waves in Applied Ethics* (Basingstoke: Palgrave Macmillan).

Bostrom, N., & Roache, R. (2011). Smart Policy: Cognitive Enhancement and the Public Interest. In J. Savulescu, R. ter Muelen, & G. Kahane (Eds.), *Enhancing Human Capabilities* (Oxford: Wiley-Blackwell).

Bostrom, N., & Sandberg, A. (2009a). Cognitive Enhancement: Methods, Ethics, Regulatory Challenges. *Science and Engineering Ethics, 15*(3), 311–341.

Bostrom, N., & Sandberg, A. (2009b). The Wisdom of Nature: An Evolutionary Heuristic for Human Enhancement. In J. Savulescu & N. Bostrom (Eds.), *Human Enhancement* (Oxford: Oxford University Press).

Bostrom, N., & Savulescu, J. (2009). Human Enhancement Ethics: The State of the Debate. In J. Savulescu & N. Bostrom (Eds.), *Human Enhancement* (Oxford: Oxford University Press).

Bouamrani, A., Peletier, L., Ratel, D., Issartel, J. P., Wion, D., Caillat, P., et al. (2005). Ethical Issues in Brain Nanomedicine. *NanoBioTechnology, 1*(3), 271–274.

Bouchard, R. (2003). Bio-Systemics Synthesis, Science and Technology Foresight Pilot Project (STFPP). Research Report #4 (Canadian National Research Council).

Brey, P. (2009). Human Enhancement and Personal Identity. In J.-K. B. Olsen, E. Selinger, & S. Riis (Eds.), *New Waves in Philosophy of Technology* (New York: Palgrave Macmillan).

Brock, D. W. (2009). Is Selection of Children Wrong? In J. Savulescu & N. Bostrom (Eds.), *Human Enhancement* (Oxford: Oxford University Press).

Brooks, R. (2008, September 21). I, Rodney Brooks, Am a Robot. *IEEE Spectrum*, http://spectrum.ieee.org/computing/hardware/i-rodney-brooks-am-a-robot.

Bruce, D. (2007). Ethical Issues in Nanobiotechnology. *Ethical Reflections on Emerging Nanobio-Technologies* (Report on an expert Working Group on Converging Technologies for Human Functional Enhancement). NanoBio-RAISE EC FP6.

Bublitz, J. C., & Merkel, R. (2009). Autonomy and Authenticity of Enhanced Personality Traits. *Bioethics, 23*(6), 360–374.

Buchanan, A., Brock, D. W., Daniels, N., & Wikler, D. (2001). *From Chance to Choice: Genetics and Justice* (Cambridge: Cambridge University Press).

Bukatman, S. (1993). *Terminal Identity: The Virtual Subject in Postmodern Science Fiction* (Durham, NC: Duke University Press).

Bullard, L. M., Browning, E. S., Clark, V. P., Coffman, B. A., Garcia, C. M., Jung, R. E., et al. (2011). Transcranial Direct Current Stimulation's Effect on Novice versus Experienced Learning. *Experimental Brain Research, 213*(1), 9–14.

Cabrera, L. (2009a). *Nanotechnology: Beyond Human Nature?* (Saarbrücken: LAP Lambert Academic Publishing).

Cabrera, L. (2009b). Nanotechnology: Changing the Disability Paradigm. *International Journal of Disability, 8*(2), http://www.ijdcr.ca/VOL08_02/articles/cabrera .shtml.

Cabrera, L. Y., Fitz, N., & Reiner, P. B. (2014). Empirical Support for the Moral Salience of the Therapy-Enhancement Distinction in the Debate over Cognitive, Affective and Social Enhancement. *Neuroethics*, http://link.springer.com/ article/10.1007/s12152-014-9223-2.

Cabrera, L., & Weckert, J. (2012). Human Enhancement and Communication: On Meaning and Shared Understanding. *Science and Engineering Ethics, 19*(3), 1039–1056.

Cahill, M., & Balice-Gordon, R. (2005). The Ethical Consequences of Modafinil Use. *Penn Bioethics Journal, 1*(1), 1–3.

Caldera, E. O. (2008). Cognitive Enhancement and Theories of Justice: Contemplating the Malleability of Nature and Self. *Journal of Evolution and Technology, 18*(1), 116–123.

Campbell, E. J. M., Scadding, J. G., & Roberts, R. S. (1979). The Concept of Disease. *The British Medical Journal, 2*(6193), 757–762.

Campbell, F. A. K. (2001). Inciting Legal Fictions-Disability's Date with Ontology and the Ableist Body of the Law. *Griffith Law Review, 10*, 42.

Campbell, F. A. K. (2008). Refusing Able(ness): A Preliminary Conversation about Ableism. *M/C Journal, 11*(3), http://journal.media-culture.org.au/index.php/ mcjournal/article/viewArticle/46

Campbell, N., O'Driscoll, A., & Saren, M. (2010). The Posthuman: The End and the Beginning of the Human. *Journal of Consumer Behaviour, 9*, 86–101.

Camus, A. (1954). *The Rebel: An Essay on Man in Revolt* (New York: A.A. Knopf).

Canton, J. (2004). Designing the Future: NBIC Technologies and Human Performance Enhancement. *Annals of the New York Academy of Sciences, 1013*, 186–198.

Carey, J. E. (2008). *Brain Facts: A Primer on the Brain and Nervous System* (Washington, DC: Society for Neuroscience).

Carnevale, A., & Battaglia, F. (2014). A Reflexive Approach to Human Enhancement: Some Philosophical Considerations. In F. Lucivero & A. Vedder (Eds.), *Beyond Therapy v. Enhancement? Multidisciplinary Analysis of a Heated Debate* (Pisa: Pisa University Press).

Cartwright, L., & Goldfarb, B. (2006). On the Subject of Neural and Sensory Prostheses. In M. E. Smith & J. Morra (Eds.), *The Prosthetic Impulse: From a Posthuman Present to a Biocultural Future* (Cambridge, MA: MIT Press).

Cascio, J. (2006). Rethinking Ourselves. *Journal of Evolution and Technology, 15*, 87–90.

Cerruti, C., & Schlaug, G. (2008). Anodal Transcranial Direct Current Stimulation of the Prefrontal Cortex Enhances Complex Verbal Associative Thought. *Journal of Cognitive Neuroscience, 21*, 1980–1987.

Chadwick, R. (2009). Therapy, Enhancement and Improvement. In B. Gordijn & R. Chadwick (Eds.), *Medical Enhancement and Posthumanity* (Dordrecht: Springer Netherlands).

Chan, S., & Harris, J. (2008). In support of Human Enhancement. *Studies in Ethics, Law, and Technology, 1*(1), 1–3.

Chatterjee, A. (2004). Cosmetic Neurology the Controversy over Enhancing Movement, Mentation, and Mood. *Neurology, 63*(6), 968–974.

Chi, R. P., & Snyder, A. W. (2011). Facilitate Insight by Non-Invasive Brain Stimulation. *Plos One 6*(2), e16655.

Clark, A. (1998). *Being There: Putting Brain, Body, and World Together Again* (Cambridge, MA: MIT Press).

Clark, A. (2004). *Natural-Born Cyborgs* (Oxford, New York: Oxford University Press).

Clark, A. (2010). Memento's Revenge: The Extended Mind, Extended. In R. Menary, *The Extended Mind* (Cambridge, MA: MIT Press).

Clark, A., & Chalmers, D. (1998). The Extended Mind. *Analysis, 58*(1), 7–19.

Clouser, K. D., Culver, C. M., & Gert, B. (1997). Malady. In J. M. Humber & R. F. Almeder (Eds.), *What Is Disease?* (Totowa, NJ: Humana Press).

Clynes, M. E., & Kline, N. (1960). Cyborgs and Space. *Astronautics, 13*, 27–31.

Coenen, C. (2014). Transhumanism and Its Genesis: The Shaping of Human Enhancement Discourse by Visions of the Future. In F. Battaglia & A. Carnevale (Eds.), *Reframing the Debate on Human Enhancement.* Humana Mente Journal of philosophical studies, Issue 26.

Coenen, C., Schuijff, M., Smits, M., Klaassen, P., Hennen, L., Rader, M., & Wolbring, G. (2009). *Human Enhancement.* European Technology Assessment Group.

Cohen Kadosh, R., Soskic, S., Iuculano, T., Kanai, R., & Walsh, V. (2010). Modulating Neuronal Activity Produces Specific and Long-Lasting Changes in Numerical Competence. *Current Biology, 20*(22), 2016–2020.

Combs, C. D. (2006). Startling Technologies Promise to Transform Medicine. *British Medical Journal, 333*(7582), 1308–1311.

Conrad, P. (1992). Medicalization and Social Control. *Annual Review of Sociology, 18*, 209–232.

CRNNI. (2006). *A Matter of Size: Triennial Review of the National Nanotechnology Initiative,* Committee to Review the National Nanotechnology Initiative (Washington, DC: National Academy Press).

CSDH. (2008). *Closing the Gap in a Generation: Health Equity through Action on the Social Determinants of Health*, Commission on Social Determinants of Health (Geneva: World Health Organization).

Daniels, N. (1985). *Just Health Care* (Cambridge: Cambridge University Press).

Daniels, N. (1992). Growth Hormone Therapy for Short Stature: Can We Support the Treatment/Enhancement Distinction. *Growth: Genetics & Hormones, 8*, 46–48.

Daniels, N. (2000). Normal Functioning and the Treatment-Enhancement Distinction. *Cambridge Quarterly of Healthcare Ethics, 9*(3), 309–322.

Daniels, N. (2008). *Just Health* (Cambridge: Cambridge University Press).

de Beauvoir, S. (2011). *The Ethics of Ambiguity* (New York: Open Road Media).

Degrazia, D. (2000). Prozac, Enhancement, and Self-Creation. *The Hastings Center Report, 30*(2), 34.

Degrazia, D. (2005). Enhancement Technologies and Human Identity. *Journal of Medicine and Philosophy, 30*(3), 261–283.

Desmond, J. E., & Pascual-Leone, A. (Eds.) (2006). TMS Improvement of Human Cognitive Abilities. *Behavioral Neurology, 17*(3 & 4), 131–203.

Dick, P. K. (1978, November 8). *How to Build a Universe That Doesn't Fall Apart Two Days Later*, http://downlode.org/Etext/how_to_build.html.

Dockery, C. A., Hueckel-Weng, R., Birbaumer, N. & Plewnia, C. (2009). Enhancement of Planning Ability by Transcranial Direct Current Stimulation. *The Journal of Neuroscience, 29*(22), 7271–7277.

Donoghue, J. P. (2002). Connecting Cortex to Machines: Recent Advances in Brain Interfaces. *Nature Neuroscience, 5*(Supp), 1085–1088.

Donoghue, J. P. (2008). Bridging the Brain to the World: A Perspective on Neural Interface Systems. *Neuron, 60*(3), 511–521.

Douglas, T. (2011). Human Enhancement and Supra-Personal Moral Status. *Philosophical Studies, 162*(3), 473–497.

Drexler, E. (1986). *Engines of Creation* (New York: Anchor Books).

Drexler, E., Forrest, D., Freitas, R. A., Hall, S., Jacobson, N., et al. (2009, December 11). On Physics, Fundamentals, and Nanorobots: A Rebuttal to Smalley's Assertion That Self-Replicating Mechanical Nanorobots Are Simply Not Possible. *Institute for Molecular Manufacturing*, http://www.imm.org/publications/sciamdebate2/smalley/.

During, M. J., Young, D., Lawlor, P. A., Leone, P., & Dragunow, M. (1999). Environmental Enrichment Inhibits Spontaneous Apoptosis, Prevents Seizures and Is Neuroprotective. *Nature Medicine, 5*(4), 448–453.

Dworkin, R. (2002). Playing God: Genes, Clones and Luck. In *Sovereign Virtue: The Theory and Practice of Equality* (Cambridge, MA: Harvard University Press).

Earp, B. D., Sandberg, A., Kahane, G., & Savulescu, J. (2014). When Is Diminishment a Form of Enhancement? Rethinking the Enhancement Debate in Biomedical Ethics. *Frontiers in Systems Neuroscience, 8*(12), 1–8.

Edgar, A. (2009). The Hermeneutic Challenge of Genetic Engineering: Habermas and the Transhumanists. *Medicine, Health Care and Philosophy, 12*(2), 157–167.

Eisenberg, L. (1977). Disease and Illness Distinctions between Professional and Popular Ideas of Sickness. *Culture, Medicine and Psychiatry, 1*(1), 9–23.

Elliott, C. (1998, January 12). What's Wrong with Enhancement Technologies. *CHIPS Public Lecture*. University of Minnesota.

Elliott, C. (1999). *A Philosophical Disease* (New York: Routledge).

Elliott, C. (2004). *Better Than Well: American Medicine Meets the American Dream* (New York: Norton & Company).

Engelhardt, H. T. (1976). Is There a Philosophy of Medicine? *Proceedings of the Biennial Meeting of the Philosophy of Science Association, 2*, 94–108.

ETC Group. (2003). *The Big Down from Genes to Atoms: Technologies Converging at the Nano-Scale*, http://www.etcgroup.org/documents/TheBigDown.pdf.

ETC Group. (2005). *A Tiny Primer on Nano-Scale Technologies and 'The Little BANG Theory'*, http://www.etcgroup.org/upload/publication/55/01/tinyprimer _english.pdf.

Farah, M. J. (2004). Emerging Ethical Issues in Neuroscience. *Nature Neuroscience, 5*(11), 1123–1130.

Farah, M. J. (2010). *Neuroethics: An Introduction with Readings* (Cambridge, MA: MIT Press).

Farah, M. J., Illes, J., Cook-Deegan, R., Gardner, H., Kandel, E., King, P., et al. (2004). Neurocognitive Enhancement: What Can We Do and What Should We Do? *Nature Reviews Neuroscience, 5*(5), 421–425.

Farah, M. J., & Heberlein, A. S. (2007). Personhood and Neuroscience: Naturalizing or Nihilating? *The American Journal of Bioethics, 7*(1), 37–48.

Farah, M. J., & Wolpe, P. R. (2004). Monitoring and Manipulating Brain Function: New Neuroscience Technologies and their Ethical Implications. *Hastings Center Report, 34*(3), 35–45.

Fernández-Armesto, F. (2005). *So You Think You're Human?: A Brief History of Humankind* (Oxford: Oxford University Press).

Fertonani, A., Rosini, S., Cotelli, M., Rossini, P. M., & Miniussi, C. (2010). Naming Facilitation Induced by Transcranial Direct Current Stimulation. *Behavioural Brain Research, 208*(2), 311–318.

Feynman, R. P. (1960). There's Plenty of Room at the Bottom: An Invitation to Enter a New Field of Physics. *Engineering and Science, 23*(5), 22–36.

Floel, A., Rösser, N., Michka, O., Knecht, S., & Breitenstein, C. (2008). Noninvasive Brain Stimulation Improves Language Learning. *Journal of Cognitive Neuroscience, 20*(8), 1415–1422.

FM-2030. (1989). *Are You a Transhuman?: Monitoring and Stimulating Your Personal Rate of Growth in a Rapidly Changing World* (New York: Warner Books).

Ford, A. (2005). Ford: Humanity: The Remix. *Utne Magazine*, http://www.utne .com/2005-05-01/humanity-the-remix.aspx.

Foucault, M. (1970). *The Order of Things: An Archaeology of the Human Sciences* (New York: Pantheon Books).

Foucault, M. (1990). *The History of Sexuality*, Vol. 1 (New York: Vintage Books).

Fregni, F., Boggio, P. S., Nitsche, M., Bermpohl, F., Antal, A., Feredoes, E., et al. (2005). Anodal Transcranial Direct Current Stimulation of Prefrontal Cortex Enhances Working Memory. *Experimental Brain Research, 166*(1), 23–30.

Freitas, R. A. (1999). *Nanomedicine. Volume 1: Basic Capabilities* (Austin, TX: Landes Bioscience).

Freitas, R. A. (2005a). Current Status of Nanomedicine and Medical Nanorobotics, *Journal of Computational and Theoretical Nanoscience, 2*(1), 1–25.

Freitas, R. A. (2005b). Nanotechnology, Nanomedicine and Nanosurgery. *International Journal of Surgery, 3*(4), 243–246.

Freitas, R. A. (2007). Personal Choice in the Coming Era of Nanomedicine. In F. Allhoff, P. Lin, J. Moor, & J. Weckert (Eds.), *Nanoethics: The Ethical and Social Implications of Nanotechnology* (Hoboken, NJ: Wiley).

Fukuyama, F. (2002). *Our Posthuman Future: Consequences of the Biotechnology Revolution* (New York: Farrar, Straus and Giroux).

Fukuyama, F. (2004). Transhumanism: The World's Most Dangerous Idea. *Foreign Policy, 144*, 42–43.

Gagnon, G., Schneider, C., Grondin, S., & Blanchet, S. (2011). Enhancement of Episodic Memory in Young and Healthy Adults: A Paired-Pulse TMS Study on Encoding and Retrieval Performance. *Neuroscience Letters, 488*(2), 138–142.

Gallate, J., Chi, R., Ellwood, S., & Snyder, A. (2009). Reducing False Memories by Magnetic Pulse Stimulation. *Neuroscience Letters*, 449,151–154.

Gallup, J. L., & Sachs, J. D. (2001). The Economic Burden of Malaria. *American Journal of Tropical Medicine and Hygiene, 64*(1–2), 85–96.

Garreau, J. (2005). *Radical Evolution: The Promise and Peril of Enhancing Our Minds, Our Bodies—and What Means to Be Human* (New York: Doubleday).

Gilligan, C. (1993). *In a Different Voice: Psychological Theory and Women's Development* (London: Harvard University Press).

Gladwin, T. E., Uyl, den T. E., Fregni, F. F., & Wiers, R. W. (2012). Enhancement of Selective Attention by tDCS: Interaction with Interference in a Sternberg Task. *Neuroscience Letters*, 512, 33–37.

Glannon, W. (2006). Neuroethics. *Bioethics, 20*(1), 37–52.

Glannon, W. (2008). *Bioethics and the Brain* (New York: Oxford University Press).

Glenn, L. M., & Dvorsky, G. (2010). Dignity and Agential Realism: Human, Posthuman, and Nonhuman. *The American Journal of Bioethics, 10*(7), 57–58.

Goldman, A. I. (1993). Ethics and Cognitive Science. *Ethics, 103*(2), 337–360.

Goodin, R. E. (1986). *Protecting the Vulnerable: A Reanalysis of Our Social Responsibilities* (Chicago, IL: University of Chicago Press).

Goodman, R. (2010). Cognitive Enhancement, Cheating, and Accomplishment. *Kennedy Institute of Ethics Journal, 20*(2), 145–160.

Goorden, L., Oudheusden, M., Evers, J., & Deblonde, M. (2008). Lose One Another ... and Find One Another in Nanospace. 'Nanotechnologies for Tomorrow's Society: A Case for Reflective Action Research in Flanders (NanoSoc)'. *Nanoethics, 2*(3), 213–230.

Gordijn, B. (2005). Nanoethics: From Utopian Dreams and Apocalyptic Nightmares towards a More Balanced View. *Science and Engineering Ethics, 11*(4), 521–533.

Gordijn, B. (2006). Converging NBIC Technologies for Improving Human Performance: A Critical Assessment of the Novelty and the Prospects of the Project. *The Journal of Law, Medicine & Ethics, 34*(4), 726–732.

Gómez-Pinilla, F. (2008). Brain Foods: The Effects of Nutrients on Brain Function. *Nature Reviews Neuroscience, 9*(7), 568–578.

Grandjean, P., 2013. *Only One Chance: How Environmental Pollution Impairs Brain Development—and How to Protect the Brains of the Next Generation* (Oxford/New York: Oxford University Press).

Gray, C. H. (Ed.) (1995). *The Cyborg Handbook* (New York/London: Routledge).

Gray, C. H. (2000). *Cyborg Citizen: Politics in the Posthuman Age* (New York/London: Routledge).

Greely, H. T. (2010). Enhancing Brains: What Are We Afraid of? *Cerebrum* http://dana.org/news/cerebrum/detail.aspx?id=28786.

Greely, H., Sahakian, B., Harris, J., Kessler, R. C., Gazzaniga, M., Campbell, P., & Farah, M. J. (2008). Towards Responsible Use of Cognitive-Enhancing Drugs by the Healthy. *Nature, 456*(7223), 702–705.

Green, A. E., Munafò, M. R., DeYoung, C. G., Fossella, J. A., Fan, J., & Gray, J. R. (2008). Using Genetic Data in Cognitive Neuroscience: From Growing Pains to Genuine Insights. *Nature Reviews Neuroscience, 9*(9), 710–720.

Greene, J., & Haidt, J. (2002). How (and Where) Does Moral Judgment Work? *Trends in Cognitive Sciences, 6*(12), 517–523.

Guan, J.-S., Haggarty, S. J., Giacometti, E., Dannenberg, J.-H., Joseph, N., Gao, J., et al. (2009). HDAC2 Negatively Regulates Memory Formation and Synaptic Plasticity. *Nature, 459*(7243), 55–60.

Habermas, J. (1985). *The Theory of Communicative Action: Lifeworld and System: A Critique of Functionalist Reason* (Boston, MA: Beacon Press).

Habermas, J. (2000). *On the Pragmatics of Communication* (Cambridge, MA: MIT Press).

Habermas, J. (2003). *The Future of Human Nature* (Cambridge: Polity).

Haggard, P., & Longo, M. R. (2010, September 8). You Are What You Touch: How Tool Use Changed the Brain's Representations of the Body. *Scientific American*, http://www.scientificamerican.com/article.cfm?id=you-are-what-you-touch.

Haidt, J. (2001). The Emotional Dog and Its Rational Tail: A Social Intuitionist Approach to Moral Judgment. *Psychological Review, 108*(4), 814.

Halberstam, J. M., & Livingston, I. (1995). *Posthuman Bodies* (Bloomington, IN: Indiana University Press).

Haldane, J. B. S. (1923, February 4). Daedalous or Science and the Future. *A Paper Read to the Heretics Cambridge* (New York: Dutton & Company).

Hamani, C., McAndrews, M. P., Cohn, M., Oh, M., Zumsteg, D., Shapiro, C. M., et al. (2008). Memory Enhancement Induced by Hypothalamic/Fornix Deep Brain Stimulation. *Annals of Neurology, 63*(1), 119–123.

Hamilton, R., Messing, S., & Chatterjee, A. (2011). Rethinking the Thinking Cap: Ethics of Neural Enhancement Using Noninvasive Brain Stimulation. *Neurology, 76*(2), 187–193.

Hamlett, P, Cobb, M. D., & Guston, D. H. (2008). National Citizens' Technology Forum: Nanotechnologies and Human Enhancement, CNS-ASU Report #R08-0002 (The Center for Nanotechnology in Society, Arizona State University).

Haraway, D. J. (1991). A Cyborg Manifesto: Science, Technology, and Socialist-feminism in the Late Twentieth Century. In *Simians, Cyborgs and Women: The Reinvention of Human Nature* (New York: Routledge).

Harris, J. (2007). *Enhancing Evolution* (Princeton, NJ: Princeton University Press).

Harris, J. (2009). Enhancements Are a Moral Obligation. In J. Savulescu & N. Bostrom (Eds.), *Human Enhancement* (Oxford: Oxford University Press).

Harris, J. (2010). Taking the 'Human' Out of Human Rights. *Cambridge Quarterly of Healthcare Ethics, 20*(1), 9–20.

Hassan, I. (1977). Prometheus as Performer: Toward a Posthumanist Culture? In M. Benamou & C. Caramello (Eds.), *Performance in Postmodern Culture* (Madison, WI: Coda Press).

Hayles, N. K. (1999). *How We Became Posthuman* (London: University of Chicago Press).

Hayles, N. K. (2010, October 28). Wrestling with Transhumanism. Paper presented at the Transhumanism and the Meanings of Progress workshop. Tempe, AZ, http://www.metanexus.net/magazine/tabid/68/id/10543/Default.aspx.

Heidegger, M. (1982). *The Question Concerning Technology, and Other Essays* (New York: HarperCollins).

Held, V. (2001). Feminist Moral Inquiry: The Role of Experience. *Janus Head, 5*(1), http://www.janushead.org/5-1/held.cfm.

Held, V. (2006). *The Ethics of Care* (New York: Oxford University Press).

Hertrich, I., Dietrich, S., Moos, A., Trouvain, J., & Ackermann, H. (2009). Enhanced Speech Perception Capabilities in a Blind Listener Are Associated with Activation of Fusiform Gyrus and Primary Visual Cortex. *Neurocase, 15*(2), 163–170.

Hescham, S., Lim, L. W., Jahanshahi, A., Steinbusch, H. W. M., Prickaerts, J., Blokland, A. & Temel, Y. (2012). Deep Brain Stimulation of the Forniceal Area Enhances Memory Functions in Experimental Dementia: The Role of Stimulation Parameters. *Brain Stimulation, 6*, 72–77.

Hilgetag, C. C., Théoret, H., & Pascual-Leone, A. (2001). Enhanced Visual Spatial Attention Ipsilateral to rTMS-induced 'Virtual Lesions' of Human Parietal Cortex. *Nature Neuroscience, 4*(9), 953–957.

Hippocrates. (n.d.). *On the Sacred Diseased*. The Internet Classic Archive. http://classics.mit.edu/Hippocrates/sacred.html.

Hodgins, D., Bertsch, A., Post, N., Frischholz, M., Volckaerts, B., Spensley, J., et al. (2008). Healthy Aims: Developing New Medical Implants and Diagnostic Equipment. *Pervasive Computing, IEEE, 7*(1), 14–21.

Hoerl, C. (1999). Memory, Amnesia and the Past. *Mind & Language, 14*(2), 227–251.

Hofmann, B. (2001). The Technological Invention of Disease. *Medical Humanities, 27*(1), 10–19.

Holm, S. (2013). Does Nanotechnology Require a New 'Nanoethics?' In B. Gordijn & A. M. Cutter (Eds.), *In Pursuit of Nanoethics* (Dordrecht: Springer Netherlands).

Hook, C. C. (2007). Nanotechnology and the Future of Medicine. In N. Cameron & E. Mitchell (Eds.), *Nanoscale: Issues and Perspectives for the Nano Century* (Hoboken, NJ: Wiley).

Hopkins, P. D. (2008). A Moral Vision of Transhumanism. *Journal of Evolution and Technology, 19*, 3–7.

Hu, R., Eskandar, E., & Williams, Z. (2009). Role of Deep Brain Stimulation in Modulating Memory Formation and Recall. *Neurosurgical Focus, 27*(1), E3.

Hughes, J. (2004). *Citizen Cyborg: Why Democratic Societies Must Respond to the Redesigned Human of the Future* (Boulder, CO: Westview Press Books).

Hughes, J. (2010a). Technoprogressive Biopolitics and Human Enhancement. In Jonathan D. Moreno & Sam Berger (Eds.), *Progress in Bioethics: Science, Policy, and Politics* (Cambridge, MA: MIT Press).

Hughes, J. (2010b, May 9). Beyond the Human Race – And 'Human-Racism.' *Io9 Magazine Online*, http://io9.com5534623/beyond-the-human-race-+-and-human+racism.

Hume, D. (2004). *A Treatise of Human Nature* (New York: Digireads Publishing).

Hurd, D. J. (2005). Converging Technologies in Developing Countries: Passionate Voices, Fruitful Actions. In W. S. Bainbridge & M. C. Roco (Eds.), *Managing*

Nano-Bio-Info-Cogno Innovations: Converging Technologies in Society (Dordrecht: Springer Netherlands).

Huxley, A. (1932). *Brave New World*, http://www.huxley.net/bnw/1st-edition.html

Huxley, J. (1957). Transhumanism. In *New Bottles for New Wine* (London: Chatto & Windus).

Ilieva, I. P., & Farah, M. J. (2012). Enhancement Stimulants: Perceived Motivational and Cognitive Advantages. *Frontiers in Neuroscience, 7*, 198.

Illes, J., 2006. *Neuroethics: Defining the Issues in Theory, Practice, and Policy* (Oxford: Oxford University Press).

Illich, I. (1975). *Medical Nemesis: The Expropriation of Health* (London: Calder & Boyars; New York: Pantheon Books).

Ilyas, M., & Mahgoub, I. (2006). *Smart Dust: Sensor Network Applications, Architecture and Design* (Boca Raton, FL: CRC Press).

Ingeborg, A., & Dalmo, R. A. (2007). Nanotechnologies and Risk: What Are the Issues? In F. Allhoff, P. Lin, J. Moor, & J. Weckert (Eds.), *Nanoethics: The Ethical and Societal Implications of Nanotechnology* (Hoboken, NJ: Wiley).

Irwin, A., & Scali, E. (2010). *Action on the Social Determinants of Health: Learning from Previous Experiences*. Social Determinants of Health Discussion Paper 1 (Geneva: World Health Organization).

Iuculano, T., & Kadosh, R. C. (2013). The Mental Cost of Cognitive Enhancement. *Journal of Neuroscience, 33*(10), 4482–4486.

Jacobson, N. (2012). *Dignity and Health* (Nashville, TN: Vanderbilt University Press).

Jak, A., Seelye, A. M. & Jurick, S. M. (2012). Crosswords to Computers: A Critical Review of Popular Approaches to Cognitive Enhancement. *Neuropsychology Review, 23*(1), 13–26.

Javadi, A. H., & Cheng, P. (2013). Transcranial Direct Current Stimulation (tDCS) Enhances Reconsolidation of Long-Term Memory. *Brain Stimulation, 6*(4), 668–674.

Johnson, C. (2008, December 5). Nanotubes Shown to Boost Neural Signals. *EETimes*. http://www.eetimes.com.news/latest/showArticle.jhtml?articleID=212700093.

Joshi, M., & Bhattacharyya, A. (2011). Nanotechnology: A New Route to High-Performance Functional Textiles. *Textile Progress, 43*(3), 155–233.

Jotterand, F. (2008). Beyond Therapy and Enhancement: The Alteration of Human Nature. *Nanoethics, 2*(1), 15–23.

Jotterand, F. (2010). Human Dignity and Transhumanism: Do Anthro-Technological Devices Have Moral Status? *The American Journal of Bioethics, 10*(7), 45–52.

Joy, B. (2000). Why the Future Doesn't Need Us. In F. Allhoff et al. (Eds.), *Nanoethics: The Ethical and Social Implications of Nanotechnology* (Hoboken, NJ: Wiley).

Juengst, E. T. (1997). Can Enhancement Be Distinguished from Prevention in Genetic Medicine? *Journal of Medicine and Philosophy, 22*(2), 125–142.

Juengst, E. T. (1998). What Does Enhancement Mean. In E. Parens, *Enhancing Human Traits* (Washington, DC: Georgetown University Press).

Kamm, F. M. (2009). What Is and Is Not Wrong with Enhancement? In J. Savulescu & N. Bostrom, *Human Enhancement* (Oxford: Oxford University Press).

Kandel, E. R. (2000). Neuroscience: Breaking Down Scientific Barriers to the Study of Brain and Mind. *Science, 290*(5494), 1113–1120.

Kass, L. R. (1985). *Toward a More Natural Science: Biology and Human Affairs* (London: The Free Press).

Kass, L. R. (1997). Wisdom of Repugnance: Why We Should Ban the Cloning of Humans *The Valparaiso University Law Review, 32,* 679.

Kearnes, M. B., Macnaghten, P. M., & Wilsdon, J. (2006). *Governing at the Nanoscale: People, Policies and Emerging Technologies* (London: DEMOS).

Keiper, A. (2007). Nanoethics as a Discipline. *The New Atlantis, 16,* 55–67.

Kelly, K. (2006, March 15). *Will Spiritual Robots Replace Humanity by 2100?* The Technium, http://www.kk.org/thetechnium/archives/2006/03/will_spiritual.php.

Kelly, M., Morgan, A., Bonnefoy, J., Butt, J., & Bergman, V. (2007). The Social Determinants of Health: Developing an Evidence Base for Political Action. *Measurement and Evidence Knowledge Network* (Geneva: World Health Organization).

Kennedy, D. (2003). Neuroethics: An Uncertain Future. Presented at the Program and Abstracts of the 33rd Annual Meeting of the American Society for Neuroscience, Society for Neuroscience New Orleans, LA.

Keravnou, E. T., Garbay, C., Baud, R., & Wyatt, J. (Eds.). (1997). *Artificial Intelligence in Medicine* (Berlin-Heidelberg: Springer-Verlag).

Khushf, G. (2004). Systems Theory and the Ethics of Human Enhancement: A Framework for NBIC Convergence. *Annals of the New York Academy of Sciences, 1013*(1), 124–149.

Khushf, G. (2005). The Use of Emergent Technologies for Enhancing Human Performance: Are We Prepared to Address the Ethical and Policy Issues? *Public Policy and Practice, 4*(2), 1–17.

Kim, Y.-H., Park, J.-W., Ko, M.-H., Jang, S. H., & Lee, P. K. W. (2004). Facilitative Effect of High Frequency Subthreshold Repetitive Transcranial Magnetic Stimulation on Complex Sequential Motor Learning in Humans. *Neuroscience Letters, 367*(2), 181–185.

Kirkwood, T., Bond, J., May, C., McKeith, I., & Teh, M.-M. (2009). Mental Capital through Life: Future Challenges. In C. Cooper, U. Goswami, & B. J. Sahakian (Eds.), *Mental Capital and Well-Being* (Oxford: Wiley).

Kirmayer, L. J. (1988). Mind and Body as Metaphors: Hidden Values in Biomedicine. In M. Lock & D. R. Gordon (Eds.), *Biomedicine Examined* (Dordrecht: Springer Netherlands).

Kirsch, S. (2006, November 9). Identifying Terrorists Before They Strike by Using Computerized Knowledge Assessment. *Skirsh.com,* http://www.skirsch.com/politics/plane/ultimate.htm.

Kirsh, D., & Maglio, P. (1994). On Distinguishing Epistemic from Pragmatic Action. *Cognitive Science, 18,* 513–549.

Klaming, L., & Haselager, P. (2010). Did My Brain Implant Make Me Do It? Questions Raised by DBS Regarding Psychological Continuity, Responsibility for Action and Mental Competence. *Neuroethics, 6*(3), 527–539.

Knoch, D., Pascual-Leone, A., Meyer, K., Treyer, V., & Fehr, E. (2006). Diminishing Reciprocal Fairness by Disrupting the Right Prefrontal Cortex. *Science, 314*(5800), 829–832.

Kobayashi, M., Hutchinson, S., Theoret, H., Schlaug, G., & Pascual-Leone, A. (2004). Repetitive TMS of the Motor Cortex Improves Ipsilateral Sequential Simple Finger Movements. *Neurology, 62*(1), 91–98.

Kosfeld, M., Heinrichs, M., Zak, P. J., Fischbacher, U., & Fehr, E. (2005). Oxytocin Increases Trust in Humans. *Nature Cell Biology, 435*(7042), 673–676.

Kramer, P. D. (1993). *Listening to Prozac* (New York: Viking Penguin).

Kranzberg, M. (1986). Technology and History: 'Kranzberg's Laws.' *Technology and Culture, 27*, 544–560.

Kuhn, T. S. (2012). *The Structure of Scientific Revolutions*, 4th edn (Chicago, IL: University of Chicago Press).

Kuniavsky, M., 2010. *Smart Things: Ubiquitous Computing User Experience Design* (Burlington, MA: Elsevier).

Kurzweil, R. (2005). *The Singularity Is Near* (New York: Penguin).

Kurzweil, R. (2007). On the National Agenda: U.S. Congressional Testimony on the Societal Implications of Nanotechnology. In F. Allhoff, P. Lin, J. Moor, & J. Weckert (Eds.), *Nanoethics: The Ethical and Social Implications of Nanotechnology* (Hoboken, NJ: Wiley).

Larriviere, D., Williams, M. A., Rizzo, M., & Bonnie, R. J. (2009). Responding to Requests from Adult Patients for Neuroenhancements: Guidance of the Ethics, Law and Humanities Committee. *Neurology, 73*(17), 1406–1412.

LeDoux, J. (2003). *Synaptic Self: How Our Brains Become Who We Are* (New York: Penguin).

Lee, H. J., & Ho, W. (1999). Single-Bond Formation and Characterization with a Scanning Tunneling Microscope. *Science, 286*(5445), 1719–1722.

Lenay, C., Gapenne, O., Hanneton, S., & Marque, C. (2003). Sensory Substitution: Limits and Perspective. In Y. Hatwell, S. Arlette, & E. Gentaz (Eds.), *Touching for Knowing: Cognitive Psychology of Haptic Manual Perception* (Amsterdam: John Benjamins).

Levin, Y. (2003). The Paradox of Conservative Bioethics. *New Atlantis, 1*(Spring), 53–65.

Levy, N. (2007). *Neuroethics: Challenges for the 21st Century* (Cambridge: Cambridge University Press).

Levy, N. (2012). Ecological Engineering: Reshaping Our Environments to Achieve Our Goals. *Philosophy & Technology, 25*(4), 589–604.

Lewis, S. C. (1943). *The Abolition of Man* (Québec, QC: Samizdat University Press).

Liao, S. M., & Sandberg, A. (2008). The Normativity of Memory Modification. *Neuroethics, 1*(2), 85–99.

Lin, P. (2007). Nanotechnology Bound: Evaluating the Case for More Regulation. *Nanoethics, 1*(2), 105–122.

Lin, P., Abney, K., & Bekey, G. A. (2012). Robot Ethics: The Ethical and Social Implications of Robotics (Cambridge, MA: MIT Press).

Lin, P., & Allhoff, F. (2006). Nanoethics and Human Enhancement: A Critical Evaluation of Recent Arguments. *Nanotechnology Perceptions, 2*(1), 47.

Lin, P., & Allhoff, F. (2008a). Against Unrestricted Human Enhancement. *Journal of Evolution and Technology, 18*(1), 35.

Lin, P., & Allhoff, F. (2008b). Untangling the Debate: The Ethics of Human Enhancement. *Nanoethics, 2*(3), 251–264.

Lindsay, R. (2005). Enhancements and Justice: Problems in Determining the Requirements of Justice in a Genetically Transformed Society. *Kennedy Institute of Ethics Journal, 15*(1), 3–38.

Lo, Y. L., Fook-Chong, S., & Tan, E. K. (2003). Increased Cortical Excitability in Human Deception. *NeuroReport, 14*(7), 1021–1024.

Locke, J. (1768). *An Essay Concerning Human Understanding* (London: H. Woodfall).

Lovelock, J. E., & Margulis, L. (1974). Atmospheric Homeostasis by and for the Biosphere: The Gaia Hypothesis. *Tellus, 26*(1–2), 2–10.

Luber, B., Fisher, C., Appelbaum, P. S., Ploesser, M., & Lisanby, S. H. (2009). Non-invasive Brain Stimulation in the Detection of Deception: Scientific Challenges and Ethical Consequences. *Behavioral Sciences & the Law, 27*(2), 191–208.

Luber, B., & Lisanby, S. H. (2014). Enhancement of Human Cognitive Performance Using Transcranial Magnetic Stimulation (TMS). *NeuroImage, 85*, 961–970.

Lynch, Z. (2004). Neurotechnology and Society (2010–2060). *Annals of the New York Academy of Sciences, 1013*(1), 229–233.

Lynch, Z. (2005). Neuropolicy (2005–2035): Converging Technologies Enable Neurotechnology Creating New Ethical Dilemmas. In W. S. Bainbridge & M. C. Roco (Eds.), *Managing Nano-Bio-Infocogno Innovations: Converging Technologies in Society* (Dordrecht: Springer Netherlands).

MacCormick, N. (1996). Justice as Impartiality: Assenting with Anti-contractualist Reservations. *Political Studies, 44*(2), 305–310.

MacKenzie, D. A., & Wajcman, J. (1985). *The Social Shaping of Technology: How the Refrigerator Got Its Hum* (Philadelphia, PA: Open University Press).

Mader, K., Brune, H., Ernst, H., Grunwald, A., Grünwald, W., Hofmann, H., et al. (2006). *Nanotechnology: Assessment and Perspectives* (Berlin Heidelberg: Springer-Verlag).

Malsch, I., & Nielsen, K. H. (2010). Nanobioethics: ObservatoryNano, 2nd Annual Report on Ethical and Societal Aspects of Nanotechnology, http://www.observatorynano.eu/project/document/2673/.

Marklund, M., Brodin, G., Stenflo, L., & Liu, C. S. (2008). New Quantum Limits in Plasmonic Devices. *Europhysics Letters, 84*(1), 17006.

Marshall, D. (2010). *The Ipswich Study: A Review of Longitudinal Methodology.* Healthy Communities Research Centre, The University of Queensland.

Marshall, L. (2004). Transcranial Direct Current Stimulation during Sleep Improves Declarative Memory. *Journal of Neuroscience, 24*(44), 9985–9992.

Maslen, H. (2014). Pharmacological Cognitive Enhancement—How Neuroscientific Research Could Advance Ethical Debate. *Frontiers in Human Neuroscience, 8*, 953.

Mazlish, B. (1967). The Fourth Discontinuity. *Technology and Culture, 8*, 1–15.

McKenny, G. P. (1997). *To Relieve the Human Condition: Bioethics, Technology, and the Body* (Albany: State University of New York Press).

Miah, A. (2003). Be Very Afraid: Cyborg Athletes, Transhuman Ideals and Posthumanity. *Journal of Evolution and Technology, 13*(2), http://jetpress.org/volume13/miah.htm.

Miah, A. (2008). Justifying Human Enhancement: The Accumulation of Biocultural Capital. *Perspectives on Our Ethical Future: Boundaries to Human Enhancements* (London: Royal Society for the Encouragement of Arts, Manufactures and Commerce).

Miah, A. (2009). A Critical History of Posthumanism. In A. M. Cutter, B. Gordijn, G. E. Marchant, A. Pompidou, & R. Chadwick, *Medical Enhancement and Posthumanity* (Dordrecht: Springer Netherlands).

Milburn, C. (2002). Nanotechnology in the Age of Posthuman Engineering: Science Fiction as Science. *Configurations, 10*(2), 261–295.

Mill, J. S. (1863). *On Liberty* (Boston: Ticknor and Fields).

Miller, D. D. (2010). Food Nanotechnology: New Leverage against Iron Deficiency. *Nature Nanotechnology, 5*(5), 318–319.

Miller, P., Parker, S., & Gillinson, S. (2004). *Disablism: How to Tackle the Last Prejudice* (London: DEMOS).

Miller, P., & Wilsdon, J. (2006). *Better Humans? The Politics of Human Enhancement and Life Extension* (London: DEMOS).

Mnyusiwalla, A., Daar, A. S., & Singer, P. A. (2003). 'Mind the Gap': Science and Ethics in Nanotechnology. *Nanotechnology, 14*(3), R9.

Moor, J., & Weckert, J. (2004). Nanoethics: Assessing the Nanoscale from an Ethical Point of View. In D. Baird, A. Nordmann, & J. Schummer (Eds.), *Discovering the Nanoscale* (Amsterdam: IOS Press).

Mora, F., Segovia, G., & del Arco, A. (2007). Aging, Plasticity and Environmental Enrichment: Structural Changes and Neurotransmitter Dynamics in Several Areas of the Brain. *Brain Research Reviews, 55*(1), 78–88.

Mottaghy, F. M., Hungs, M., Brügmann, M., Sparing, R., Boroojerdi, B., Foltys, H., et al. (1999). Facilitation of Picture Naming after Repetitive Transcranial Magnetic Stimulation. *Neurology, 53*(8), 1806–1806.

Mottaghy, F. M., Sparing, R., & Töpper, R. (2006). Enhancing Picture Naming with Transcranial Magnetic Stimulation. *Behavioural Neurology, 17*(3–4), 177–186.

Moynihan, R., Heath, I., & Henry, D. (2002). Selling Sickness: The Pharmaceutical Industry and Disease Mongering. *British Medical Journal, 324*(7342), 886–891.

Mulhall, D. (2002). *Our Molecular Future: How Nanotechnology, Robotics, Genetics and Artificial Intelligence Will Transform Our World* (Amherst, NY: Prometheus Books).

Munkittrick, K. (2010, March 29). Transhumanism and Superheroes. *H+Magazine*, http://hplusmagazine.com/2010/03/29/transhumanism-and-superheroes.

Naam, R. (2005). *More Than Human* (New York: Broadway Books).

Nagel, T. (1974). What Is It Like to Be a Bat? *The Philosophical Review, 83*(4), 435–450.

National Research Council. (2008). *Emerging Cognitive Neuroscience and Related Technologies* (Washington, DC: National Academies Press).

National Research Council. (2009). *Persistent Forecasting of Disruptive Technologies* (Washington, DC: National Academies Press).

Neurotechnology Industry Organisation. Bill for a National Neurotechnology Initiative. 2d session: S2989. Retrieved from http://www.opencongress.org/bill/110h5989/text.

Nichter, M. (2008). *Global Health: Why Cultural Perceptions, Social Representations, and Biopolitics Matter* (Tucson, AZ: University of Arizona Press).

Nicolelis, M. A. (2003). Brain-Machine Interfaces to Restore Motor Function and Probe Neural Circuits. *Nature Reviews Neuroscience, 4*(5), 417–422.

Nietzsche, F. (1968). *The Will to Power* (New York: Vintage Books).

Nietzsche, F. (1999). *Thus Spake Zarathustra: A Book for All and None* (Project Gutenberg).

Nietzsche, F. (2001). *Nietzsche: The Gay Science* (Cambridge: Cambridge University Press).

Nordenfelt, L. (2003). An Evolutionary Concept of Health: Health as a Natural Function. In L. Nordenfelt & P.-E. Liss (Eds.), *Dimensions of Health and Health Promotion* (Amsterdam, New York: Editions Rodopi).

Nordmann, A. (Rapporteur) (2004). *Converging Technologies: Shaping the Future of European Societies,* Report from the High Level Expert Group on 'Foresighting the New Technology Wave' (Luxembourg: European Commission Research).

Nordmann, A. (2007). If and Then: A Critique of Speculative NanoEthics. *Nanoethics, 1*(1), 31–46.

NSTC. (2002). *National Nanotechnology Initiative: The Initiative and Its Implementation Plan* (Washington, DC: National Science and Technology Council).

Nussbaum, M. C. (2001). *Women and Human Development: The Capability Approach* (Cambridge: Cambridge University Press).

Nussbaum, M., & Sen, A. (Eds.) (1993). *The Quality of Life* (New York: Clarendon Press).

Ohn, S. H., Park, C.-I., Yoo, W.-K., Ko, M.-H., Choi, K. P., Kim, G.-M., et al. (2008). Time-Dependent Effect of Transcranial Direct Current Stimulation on the Enhancement of Working Memory. *NeuroReport, 19*(1), 43–47.

Olivier, P., Wherton, J., & Monk, A. (2009). Technology Solutions to Prevent Waste of Mental Capital, In C. Cooper, U. Goswami, & B. J. Sahakian (Eds.), *Mental Capital and Well-Being* (Oxford: Wiley).

O'Mathuna, D. (2009). *Nanoethics: Big Ethical Issues with Small Technology (think now)* (London: Continuum).

O'Neill, J. (2014). Antimicrobial Resistance: Tackling a Crisis for the Health and Wealth of Nations, http://www.his.org.uk/files/4514/1829/6668/AMR_Review _Paper_-_Tackling_a_crisis_for_the_health_and_wealth_of_nations_1.pdf.

Oriá, R. B., Patrick, P. D., Zhang, H., Lorntz, B., de Castro Costa, C. M., Brito, G. A. C., et al. (2005). APOE4 Protects the Cognitive Development in Children with Heavy Diarrhea Burdens in Northeast Brazil. *Pediatric Research, 57*(2), 310–316.

Parens, E. (1998). Special Supplement: Is Better Always Good? The Enhancement Project. *The Hastings Center Report, 28*(1), S1.

Parens, E. (2005). Authenticity and Ambivalence: Toward Understanding the Enhancement Debate. *The Hastings Center Report, 35*(3), 34–41.

Parens, E. (2006). Creativity, Gratitude, and the Enhancement Debate. In J. Illes (Ed.), *Neuroethics: Defining the Issues in Theory, Practice and Policy* (Oxford: Oxford University Press).

Parens, E. (2013). On Good and Bad Forms of Medicalization. *Bioethics, 27*(1), 28–35.

Parfit, D. (1971). Personal Identity. *The Philosophical Review, 80*(1), 3–27.

PCB. (2003). *Beyond Therapy: Biotechnology and the Pursuit of Happiness.* President's Council on Bioethics (New York: HarperCollins).

PCSBI. (2014). *Gray Matters: Integrative Approaches for Neuroscience, Ethics, and Society* (Washington, DC: Presidential Commission for the Study of Bioethical Issues).

Pepperell, R. (2004). *The Posthuman Condition: Consciousness Beyond the Brain* (Bristol: Intellect Books).

Pepperell, R. (2005). Posthumans and Extended Experience. *Journal of Evolution and Technology, 14*(1), 27–41.

Persson, I., & Savulescu, J. (2008). The Perils of Cognitive Enhancement and the Urgent Imperative to Enhance the Moral Character of Humanity. *Journal of Applied Philosophy, 25*(3), 162–177.

Peters, T. (2007). Are We Playing God with Nanoenhancements? In F. Allhoff, P. Lin, J. Moor, & J. Weckert (Eds.), *Nanoethics: The Ethical and Social Implications of Nanotechnology* (Hoboken, NJ: Wiley).

Plessner, H. (1961). *Laughing and Crying: A Study of the Limits of Human Behavior.* Translated by J. S. Churchill & M. Greende (Evanston, IL: Northwestern University Press).

Pogge, T. W. (2002). Can the Capability Approach Be Justified? *Philosophical Topics, 30*(2), 167–228.

Popper, K. R. (1961). *The Poverty of Historicism* (London: Routledge & Kegan Paul).

Postman, N. (1998, March 27). *Five Things We Need to Know about Technological Change.* Conference Proceedings 'The New Technologies and the Human Person: Communicating the Faith in the New Millennium', Denver, CO.

Postman, N. (2006). *Amusing Ourselves to Death: Public Discourse in the Age of Show Business* (New York: Penguin Books).

Prüss-Üstün, A., & Corvalán, C. (2006). *Preventing Disease through Healthy Environments* (Geneva: World Health Organization).

Quibria, M. G., Ahmed, S. N., Tschang, T., & Reyes-Macasaquit, M. (2002). *Digital Divide: Determinants and Policies with Special Reference to Asia* (Manila: Asian Development Bank).

Rabinow, P., & Rose, N. (2006). Biopower Today. *BioSocieties, 1*(2), 195–217.

Racine, E. (2010). *Pragmatic Neuroethics: Improving Treatment and Understanding of the Mind-Brain* (Cambridge, MA: MIT Press).

Racine, E., & Forlini, C. (2010). Cognitive Enhancement, Lifestyle Choice or Misuse of Prescription Drugs? *Neuroethics, 3*(1), 1–4.

Rampon, C., Jiang, C. H., Dong, H., Tang, Y. P., Lockhart, D. J., Schultz, P. G., et al. (2000). Effects of Environmental Enrichment on Gene Expression in the Brain. *Proceedings of the National Academy of Sciences, 97*(23), 12880–12884.

Ratner, M. A., & Ratner, D. (2003). *Nanotechnology: A Gentle Introduction to the Next Big Idea* (Upper Saddle River, NJ: Prentice Hall Professional).

Ravelingien, A., Braeckman, J., Crevits, L., De Ridder, D., & Mortier, E. (2009). 'Cosmetic Neurology' and the Moral Complicity Argument. *Neuroethics, 2*(3), 151–162.

Rawls, J. (1971). *A Theory of Justice* (Cambridge, MA: Harvard University Press).

Rawls, J. (2001). *Justice as Fairness,* 2nd edn (Cambridge, MA: Harvard University Press).

Repantis, D., Schlattmann, P., Laisney, O., & Heuser, I. (2010). Modafinil and Methylphenidate for Neuroenhancement in Healthy Individuals: A Systematic Review. *Pharmacological Research, 62*(3), 187–206.

Roache, R. (2008). Ethics, Speculation, and Values. *Nanoethics, 2*(3), 317–327.

Roache, R., & Clarke, S. (2009). Bioconservatism, Bioliberalism, and the Wisdom of Reflecting on Repugnance. *Monash Bioethics Review, 28*(4), 1–21.

Robeyns, I. (2006). The Capability Approach in Practice*. *Journal of Political Philosophy, 14*(3), 351–376.

Roco, M. C., & Bainbridge, W. S. (Eds.) (2002). *Converging Technologies for Improving Human Performance: Nanotechnology, Biotechnology, Information Technology and Cognitive Science* (Arlington, VA: NSF/Department of Commerce Arlington).

Roco, M. C., & Bainbridge, W. S. (Eds.) (2007). *Nanotechnology: Societal Implications II- Individual Perspectives* (Dordrecht: Springer Netherlands).

Roco, M. C., Bainbridge, W. S., Tonn, B., & Whitesides, G. (2013). *Convergence of Knowledge, Technology and Society: Beyond Convergence of Nano-bio-info-cognitive Technologies* (Cham, Heidelberg, New York, Dordrecht, London: Springer International Publishing).

Roco, M. C. Williams, R. S., & Alivisatos, P. (2000). *National Research Directions* (Dordrecht: Kluwer Academic Publishers).

Roden, D. (2010). Deconstruction and Excision in Philosophical Posthumanism. *The Journal of Evolution and Technology, 21*(1), 27–36.

Rose, N. (2007). *The Politics of Life Itself: Biomedicine, Power, and Subjectivity in the Twenty-First Century* (Princeton, NJ: Princeton University Press).

Roskies, A. (2010). Neuroethics: Considering Its Scope and Limits. *AJOB Neuroscience, 1*(4), 1–2.

Royal Society. (2004). *Nanoscience and Nanotechnologies* (London: Royal Society and Royal Academy of Engineering).

Ruder, W. C., Lu, T., & Collins, J. J. (2011). Synthetic Biology Moving into the Clinic. *Science, 333*(6047), 1248–1252.

Sabin, J. E., & Daniels, N. (1994). Determining 'Medical Necessity' in Mental Health Practice. *The Hastings Center Report, 24*(6), 5–13.

Sade, R. M., & Khushf, G. (1998). Gene Therapy: Ethical and Social Issues. *Journal-South Carolina Medical Association, 94*(9), 406–410.

Safire, W. (2002). Visions for a New Field of Neuroethics. In *Neuroethics: Mapping the Field Conference Proceedings* (New York: Dana Press).

Sandberg, A. (2003). Morphological Freedom: Why We Not Just Want It, But Need It. *Eudoxa Policy Studies*, http://www.eudoxa.se/content/archives/2003/10/eudoxa_policy_s_3.html.

Sandberg, A., & Bostrom, N. (2008). *Whole Brain Emulation: A Roadmap*, Future of Humanity Institute, Oxford University, http://www.fhi.ox.ac.uk/reports/2008-3.pdf.

Sandel, M. (2004, January 11). The Case against Perfection. *The Atlantic Monthly*, 1–10.

Sandler, R. (2009). Nanomedicine and Nanomedical Ethics. *The American Journal of Bioethics, 9*(10), 16–17.

Sandler, R. L. (2014). *Ethics and Emerging Technologies* (Boston, MA: Palgrave Macmillan).

Santayana, G. (2005). *The Life of Reason the Phases of Human Progress* (Project Gutenberg).

Sarewitz, D., & Karas, T. (2006). *Policy Implications of Technologies for Cognitive Enhancement*, http://www.cspo.org/documents/FinalEnhancedCognition Report.pdf.

Sartre, J. P. (2012). *Being and Nothingness* (New York: Open Road Media).

Savulescu, J. (2006). Justice, Fairness, and Enhancement. *Annals of the New York Academy of Sciences, 1093*(1), 321–338.

Savulescu, J. (2009). The Human Prejudice and the Moral Status of Enhanced Beings: What Do We Owe the Gods? In N. Bostrom & J. Savulescu (Eds.), *Human Enhancement* (Oxford: Oxford University Press).

Savulescu, J., Meulen, ter R., & Kahane, G. (2011). *Enhancing Human Capacities* (Oxford: Wiley-Blackwell).

Sawyer, R. J. (2007). Robot Ethics. *Science, 318*(5853), 1037–1037.

Schechtman, M. (2007). *The Constitution of Selves* (Ithaca, NY: Cornell University Press).

Scheper-Hughes, N., & Lock, M. M. (1986). Speaking 'Truth' to Illness: Metaphors, Reification, and a Pedagogy for Patients. *Medical Anthropology Quarterly, 17*(5), 137–140.

Scheper-Hughes, N., & Lock, M. M. (1987). The Mindful Body: A Prolegomenon to Future Work in Medical Anthropology. *Medical Anthropology Quarterly, 1*(1), 6–41.

Schermer, M. (2008). Enhancements, Easy Shortcuts, and the Richness of Human Activities. *Bioethics, 22*(7), 355–363.

Schermer, M. (2011). Health, Happiness and Human Enhancement—Dealing with Unexpected Effects of Deep Brain Stimulation. *Neuroethics, 6*(3), 435–445.

Schermer, M., Bolt, I., de Jongh, R., & Olivier, B. (2009). The Future of Psychopharmacological Enhancements: Expectations and Policies. *Neuroethics, 2*, 75–87.

Schneider, S. (2010). *Science Fiction and Philosophy: From Time Travel and Superintelligence* (West Sussex: Wiley-Blackwell).

Scholey, A. B., Moss, M. C., & Wesnes, K. (1998). Oxygen and Cognitive Performance: The Temporal Relationship between Hyperoxia and Enhanced Memory. *Psychopharmacology, 140*(1), 123–126.

Schummer, J. (2004). 'Societal and Ethical Implications of Nanotechnology': Meanings, Interest Groups, and Social Dynamics. *Techné: Research in Philosophy and Technology, 8*(2), http://scholar.lib.vt.edu/ejournals/SPT/v8n2/schummer .html.

Selgelid, M. J. (2007). An Argument against Arguments for Enhancement. *Studies in Ethics, Law, and Technology, 1*(1), art. 12.

Selgelid, M. J. (2009). Afterword: Advancing Posthuman Enhancement Dialogue. In A. M. Cutter, B. Gordijn, G. E. Marchant, A. Pompidou, & R. Chadwick (Eds.), *Medical Enhancement and Posthumanity* (Dordrecht: Springer Netherlands).

Sen, A. (1979, May 22). Equality of What? *The Tanner Lecture on Human Values* (Stanford, CA: Stanford University Press).

Silvers, A. (2008). The Right Not to Be Normal as the Essence of Freedom. *Journal of Evolution and Technology, 18*(1), 79–85.

Singer, P. (1975). *Animal Liberation* (New York: Random House).

Sismondo, S. (1993). Some Social Constructions. *Social Studies of Science, 23*(3), 515–553.

Slote, M. (2007). *The Ethics of Care and Empathy* (New York: Routledge).

Smalley, R. E. (2001). Of Chemistry, Love and Nanobots. *Scientific American, 285*(3), 76–77.

Smith, K. M. (2005). Saving Humanity?: Counter-Arguing Posthuman Enhancement. *Journal of Evolution and Technology, 14*(1), 43–53.

Smith, M., & Morra, J. (2006). *The Prosthetic Impulse: From a Posthuman Present to a Biocultural Future* (Cambridge, MA: The MIT Press).

Snyder, A. W. (2009). Explaining and Inducing Savant Skills: Privileged Access to Lower Level, Less-Processed Information. *Philosophical Transactions: Biological Sciences, 364*(1522), 1399–1406.

Snyder, A. W., Ellwood, S., & Chi, R. (2012, November 29). Switching on Creativity. *Scientific American*, http://www.scientificamerican.com/article/switching -on-creativity-special/.

Snyder, A. W., Mulcahy, E., Taylor, J. L., Mitchell, D. J., Sachdev, P., & Gandevia, S. C. (2003). Savant-like Skills Exposed in Normal People by Suppressing the Left Fronto-Temporal Lobe. *Journal of Integrative Neuroscience, 2*, 149–158.

Society for Neuroscience. (2005, November 14). New Research Finds That the Common Arguments against the Development of Cognitive Enhancers Are Misguided. *Society for Neuroscience.* http://www.sfn.org/Press-Room/News -Release-Archives/2005/.

Sorgner, S. L. (2009). Nietzsche, the Overhuman, and Transhumanism. *Journal of Evolution and Technology, 20*(1), 29–42.

Stewart, W. J. (2009). Technology Futures. In C. Cooper, U. Goswami, & B. J. Sahakian (Eds.), *Mental Capital and Well-Being* (Oxford: Wiley).

Sullivan, M. (1986). In What Sense Is Contemporary Medicine Dualistic? *Culture, Medicine and Psychiatry, 10*, 331–350.

Suthana, N., Haneef, Z., Stern, J., Mukamel, R., Behnke, E., Knowlton, B., & Fried, I. (2012). Memory Enhancement and Deep-Brain Stimulation of the Entorhinal Area. *New England Journal of Medicine, 366*(6), 502–510.

Swierstra, T., & Rip, A. (2007). Nano-Ethics as NEST-Ethics: Patterns of Moral Argumentation about New and Emerging Science and Technology. *Nanoethics, 1*(1), 3–20.

Synofzik, M., Schlaepfer, T. E., & Fins, J. J. (2012). How Happy Is Too Happy? Euphoria, Neuroethics, and Deep Brain Stimulation of the Nucleus Accumbens. *AJOB Neuroscience, 3*(1), 30–36.

Tamburrini, G. (2009). Robot Ethics: A View from the Philosophy of Science. In R. Capurro & M. Nagenborg (Eds.), *Ethics and Robotics* (Heidelberg: IOS Press).

Taniguchi, N. (1974). On the Basic Concept of 'Nano-Technology'. *Proceedings of International Conference on Production Engineering Part 2* (Tokyo: Japan Society of Precision Engineering).

Task group summaries. (2007). *NAKFI Smart Prosthetics: Exploring Assistive Devices for the Body and Mind* (Washington, DC: National Academy of Sciences).

Taylor, C. (1985). Theories of Meaning. In *Human Agency and Language. Philosophical Papers 1* (Cambridge: Cambridge University Press).

Taylor, C. (1991). *The Ethics of Authenticity* (Cambridge, MA: Harvard University Press).

Tegart, G. (2006). *Environmental, Social, Legal and Ethical Aspects of Development of Nanotechnologies in Australia* (Parkville: National Academies Forum).

Thomas, R. K. (2003). *Society and Health: Sociology for Health Professionals* (New York: Kluwer Academic/Plenum Publishers).

Tomellini, R., Faure, U., & Panzer, O. (2006). *Nanomedicine Nanotechnology for Health* (Luxembourg: European Technology Platform).

Türk, V., Knowles, H., Wallbaum, H., & Kastenholz, H. (2006). *The Future of Nanotechnology: We Need to Talk*, http://www.nanologue.net.

Turkle, S. (2011). *Life on the Screen: Identity in the Age of the Internet* (New York: Simon and Schuster).

Turner, D. C., & Sahakian, B. J. (2006). Neuroethics of Cognitive Enhancement. *BioSocieties, 1*(1), 113–123.

Tversky, A., & Kahneman, D. (1974). Judgment under Uncertainty: Heuristics and Biases, *Science, 185*, 1124–1131.

UNESCO. (2006). *The Ethics and Politics of Nanotechnology* (Paris: United Nations Educational, Scientific and Cultural Organization).

van Praag, H., Kempermann, G., & Gage, F. H. (2000). Neural Consequences of Environmental Enrichment. *Nature Reviews Neuroscience, 1*(3), 191–198.

Varela, F., Thompson, E. & Rosch, E. (1991). *The Embodied Mind: Cognitive Science and Human Experience* (Cambridge, MA: MIT Press).

Verdoux, P. (2009). Transhumanism, Progress and the Future. *Journal of Evolution and Technology, 20*(2), 49–69.

Vinge, V. (1993). *The Coming Technological Singularity: How to Survive in the Post-Human Era.* Presented at the VISION-21 symposium sponsored by NASA, Lewis Research Center and the Ohio Aerospace Institute.

Walker, M. (2008). Cognitive Enhancement and the Identity Objection. *Journal of Evolution and Technology, 18*(1), 108–115.

Walsh, V., Ellison, A., Battelli, L., & Cowey, A. (1998). Task-Specific Impairments and Enhancements Induced by Magnetic Stimulation of Human Visual Area V5. *Proceedings: Biological Sciences, 265*(1395), 537–543.

Walter, S. (2009). Locked-in Syndrome, BCI, and a Confusion about Embodied, Embedded, Extended, and Enacted Cognition. *Neuroethics, 3*(1), 61–72.

Warwick, K. (2004). *I, Cyborg* (Champaign: University of Illinois Press).

Warwick, K. (2014). The Cyborg Revolution. *Nanoethics, 8*(3), 263–273.

Weckert, J. (2007). Editorial. *Nanoethics, 1*, 1–2.

Weder, A. (2007). Diseases of Civilization: An Evolutionary Legacy. In R. Nesse (Ed.), *Evolution and Medicine: How New Applications Advance Research and Practice* (London: Henry Stewart Talks Ltd).

Weisman, O., Zagoory-Sharon, O., & Fregni, F. (2012). Oxytocin Administration to Parent Enhances Infant Physiological and Behavioral Readiness for Social Engagement. *Biological Psychiatry, 72*(12), 982–989.

Weiss, P. L., Rand, D., Katz, N., & Kizony, R. (2004). Video Capture Virtual Reality as a Flexible and Effective Rehabilitation Tool. *Journal of NeuroEngineering and Rehabilitation, 1*(1), 12.

Whitehouse, P. J., Juengst, E., Mehlman, M., & Murray, T. H. (1997). Enhancing Cognition in the Intellectually Intact. *The Hastings Center Report, 27*(3), 14.

WHO. (1946). *Constitution of the World Health Organization.* World Health Organization, http://www.who.int/governance/eb/who_constitution_en.pdf.

WHO. (2002). Reducing Risks, Promoting Healthy Life. *The World Health Report* (Geneva: World Health Organization).

WHO. (2007). *Neurological Disorders: Public Health Challenges* (Geneva: World Health Organisation).

WHO. (2008). *Integrating Mental Health into Primary Care: A Global Perspective* (Geneva: World Health Organization).

Wiener, N. (1950). *The Human Use of Human Beings: Cybernetics and Society* (London: Houghton Mifflin Company).

Williams, E. A. (2007). *Good, Better, Best: The Human Quest for Enhancement*. Summary Report of an Invitational Workshop Convened by the Scientific Freedom, Responsibility and Law Program (Washington, DC: American Association for the Advancement of Science [AAAS]).

Wilson, M., Kannangara, K., Smith, G., Simmons, M., & Raguse, B. (2002). *Nanotechnology: Basic Science and Emerging Technologies* (Sydney: Chapman & Hall/CRC).

Winner, L. (1977). *Autonomous Technology: Technics-Out-of-Control as a Theme in Political Thought* (Cambridge, MA: The MIT Press).

Wittgenstein, L. (1953). *Philosophical Investigations*. Translated by G. E. M. Anscombe (Oxford: Basil Blackwell).

Wolbring, G. (2005). *The Triangle of Enhancement Medicine, Disabled People, and the Concept of Health*: A New Challenge for HTA, Health Research, and Health Policy (Edmonton: Alberta Heritage Foundation for Medical Research).

Wolbring, G. (2006). The Unenhanced Underclass. In P. Miller & J. Wilsdon, *Better Humans? The Politics of Human Enhancement and Life Extension* (London: DEMOS).

Wolbring, G. (2008a). Is There an End to Out-Able? Is There an End to the Rat Race for Abilities? *M/C Journal, 11*(3), http://journal.media-culture.org.au/index.php/mcjournal/article/viewArticle/57.

Wolbring, G. (2008b). Why NBIC? Why Human Performance Enhancement? *Innovation: The European Journal of Social Science Research, 21*(1), 25–40.

Wolpe, P. R. (2002). Treatment, Enhancement, and the Ethics of Neurotherapeutics. *Brain and Cognition*, 50, 387–395.

Wolpe, P. R. (2010). Nanoethics: Nanononsense? In L. Zoloth & M. Flory, *The Social Scale: Nanotechnology and the Weight of Justice* (Evanston, IL: Northwestern University Press).

Wood, S., Jones, R. A. L., & Geldart, A. (2003). *The Social and Economic Challenges of Nanotechnology* (Swindon: Economic & Social Research Council).

Yee, N., & Bailenson, J. (2007). The Proteus Effect: The Effect of Transformed Self-Representation on Behavior. *Human Communication Research, 33*(3), 271–290.

Young, L., Camprodon, J. A., Hauser, M., Pascual-Leone, A., Saxe, R., & Kanwisher, N. G. (2010). Disruption of the Right Temporoparietal Junction with Transcranial Magnetic Stimulation Reduces the Role of Beliefs in Moral Judgments. *Proceedings of the National Academy of Sciences of the United States of America, 107*(15), 6753–6758.

Zonneveld, L. (2008). *Reshaping the Human Condition: Exploring Human Enhancement* (The Hague: Rathenau Institute).

Index